Biodiversity and Ecosystem Insecurity

Biodiversity and Ecosystem Insecurity

A Planet in Peril

Edited by Ahmed Djoghlaf and Felix Dodds

publishing for a sustainable future
London • Washington, DC

First published in 2011 by Earthscan

Earthscan Ltd, Dunstan House, 14a St Cross Street, London EC1N 8XA, UK

Earthscan LLC,1616 P Street, NW, Washington, DC 20036, USA

Earthscan publishes in association with the International Institute for Environment and Development
For more information on Earthscan publications, see www.earthscan.co.uk or write to earthinfo@earthscan.co.uk

ISBN: 978-1-84971-219-4 hardback
ISBN: 978-1-84971-220-0 paperback

Typeset by FiSH Books
Cover design by John Charles

A catalogue record for this book is available from the British Library

Library of Congress Cataloging-in-Publication Data
Biodiversity and ecosystem insecurity : a planet in peril / edited by Ahmed Djoghlaf and Felix Dodds.
p. cm.
Includes bibliographical references and index.
ISBN 978-1-84971-219-4 (hardback) — ISBN 978-1-84971-220-0 (pbk.) 1. Biodiversity. 2. Endangered ecosystems. 3. Climatic changes—Environmental aspects. 4. Biodiversity conservation—Government policy. I. Djoghlaf, Ahmed. II. Dodds, Felix.
QH541.15.B56B5683 2011
333.95'16—dc22 2010046142

At Earthscan we strive to minimize our environmental impacts and carbon footprint through reducing waste, recycling and offsetting our CO_2 emissions, including those created through publication of this book. For more details of our environmental policy, see www.earthscan.co.uk.

Printed and bound in the UK by CPI Antony Rowe, Chippenham.
The paper used is FSC certified.

Contents

List of Figures and Tables

Figures

Tables

Foreword

Achim Steiner
United Nations Under-Secretary-General
Executive Director, United Nations Environment
Programme (UNEP)

The fuel and food crises; the ongoing financial and economic challenges; and the emerging threats of climate change and natural resource scarcity are all becoming defining issues of our time.

How the world responds will in large part define sustainability in the 21st century and the ability of 6 billion people, rising to 9 billion by 2050, to thrive let alone survive over the coming decades.

Biodiversity and ecosystems – from forests and freshwater to coral reefs, soils and even the atmosphere – are in the front line in terms of impacts but also solutions.

Feeding the world and fighting climate change, for example, will be increasingly linked with the way these nature-based assets are intelligently managed and rehabilitated or increasingly degraded and destroyed.

To date the international response has failed to match the pace of environmental change. But that is no alibi for inaction, and decisions taken at the Tenth Meeting of the Conference of the Parties (COP) of the Convention on Biological Diversity (CBD) in Nagoya have marked some important steps forward.

Encouraging decisions were also taken in Busan, Republic of Korea, last year when governments gave the green light to an Intergovernmental Science-Policy Platform on Biodiversity and Ecosystem Services (IPBES).

Meanwhile progress is happening under the United Nations (UN) climate convention on Reducing Emissions from Deforestation and Forest Degradation (REDD) in terms of triggering further investments that could represent the biggest financial injection in sustainable forestry management ever.

Action on so called non-carbon-dioxide pollutants such as black carbon, tropospheric ozone and nitrogen compounds may also prove a fruitful climate-friendly avenue with benefits to ecosystems including fisheries currently at risk from deoxygenated 'dead zones'.

Nature offers further opportunities in terms of the carbon sequestration potential of its soils and vegetation. By some estimates coastal ecosystems are also absorbing up to half the world's transport emissions while providing cost effective sea defences and nurseries for fish.

While the international response to the challenges today's world faces remain inadequate, there is a growing recognition of the opportunities of investing in the environment and nature-based assets.

This is in part driven by rapidly evolving science and also by The Economics of Ecosystems and Biodiversity (TEEB), hosted by UNEP and a key theme driving of the organization's Green Economy initiative.

In the past, the true value of biodiversity and ecosystems was all but invisible in national and global accounts. But TEEB is bringing visibility to the economic but also social benefits of the natural environment while spotlighting smart market mechanisms and policy shifts able to capture that value.

During the 2010 Millennium Development Goals (MDGs) Summit, many countries underlined the wide-ranging role the environment can play in meeting many if not all of the eight MDGs.

Investing and reinvesting in the planet's ecological infrastructure – allied to the greening of transport and energy up to agriculture and water management – offer a dynamic path towards sustainable development by decoupling growth from humanity's ecological footprint.

Meanwhile reforming perverse or harmful subsidies, from fossil fuel ones to those linked with agriculture and fisheries, could be harnessed for investing in sustainable development and a Green Economy trajectory.

Many of these landmark events and refocusing of the sustainable development path are now leading inexorably to the Rio+20 meeting in Brazil in 2012.

When the so-called Rio conventions were agreed in 1992, the world was just glimpsing future crises. Today we are staring them in the face and perhaps grasping for the first time the multiple challenges but also multiple opportunities for action.

With each passing year, the science becomes ever more sobering, the room for manoeuver becomes tighter yet paradoxically the suite of options and opportunities climbs as we gather greater and more transformational knowledge.

A response to current and future challenges will happen either by design or by default – the next two years may well define which of these two futures wins out and whether it is a planet in peril or a planet of peace and prosperity that is handed over to the next generation.

Foreword

Ryu Matsumoto
Minister of the Environment and the Minister of State for Disaster Management, Japan

Throughout the years, momentum has increased regarding the efforts to understand the concept and importance of biodiversity and its direct effect on the well-being of humankind. Despite the efforts carried out so far, recent studies in different countries have shown that a large percentage of the public in general have not heard of the word 'biodiversity' or do not know what it means. The international community has recognized the need to increase biodiversity awareness and make sustainable use of ecosystems and increase the conservation efforts, however, much remains to be done.

In the 2002 World Summit on Sustainable Development (WSSD), the international community set the goal to achieve a significant reduction of the current rate of biodiversity loss by 2010. Eight years later, it is clear that this objective has not been met and that, in fact, we still have major biodiversity loss. In 2010, declared by the UN as the International Year of Biodiversity, we have a new opportunity to develop new targets, and a new strategy based on past experiences. Japan, serving as the host country of the tenth meeting of the Conference of the Parties (COP-10) to the Convention on Biological Diversity (CBD), in Nagoya, has made significant contributions for the protection of Biodiversity. For instance, Japan is the only country out of 193 parties to the CBD that has revised its National Biodiversity Strategy Action Plan (NBSAP) three times. Japan's Fourth NBSAP was developed in March 2010 in accordance with the Basic Act of Biodiversity that was established in 2008. The fourth NBSAP, setting a time-bound mid–long-term target for 2050 and shorter-term targets for 2020, identifies key actions on biodiversity at the national and international level by mainstreaming biodiversity in the society.

Another significant Japanese contribution has been the Japan Biodiversity Outlook, which entails a comprehensive assessment of biodiversity in Japan. Launched 10 May 2010, alongside the flagship publication of the CBD, Global Biodiversity Outlook 3 (GBO3), the report evaluated the impact of the loss of biodiversity in the past 50 years as well as the current trend of biodiversity in Japan. Regarding partnerships and outreach, Japan and the United Nations University Institute of Advanced Studies (UNU-IAS) jointly initiated the Satoyama Initiative, which aims to conserve sustainable types of human-influenced natural environments, the so-called 'socio-economic landscapes', through global partnerships for the benefit of biodiversity and human well-being.

Japan, along with many other countries, is fully committed to making its contribution so that realistic, practical and effective post-2010 biodiversity targets can be established and consequently achieved. However, there are many challenges up ahead. There is an evident need for a new global strategy for action and for governments, policy-makers, and society in general to jointly identify the adequate means to achieve post-2010 targets. The value of biodiversity for our well-being needs to be grasped and integrated in all the conservation efforts so that the reduction and subsequently the halting of biodiversity loss can happen while ensuring the most sustainable delivery of ecosystem services.

Nevertheless, many other issues need to be taken into account to complement this new global strategy. For example, the need to improve the juncture between science and governance is of noteworthy importance. Despite the past efforts to harmonize the progress of science and its increasing recognition as a key element to decision-making into environmental governance, the lack of an appropriate flow of information between these two components has obstructed the efforts of halting the loss of biodiversity. Therefore, it is essential to concentrate efforts in coordinating all stakeholders for a more effective management of biodiversity and ecosystem services. Opportunely, in June 2010, history was made when governments reached consensus on the creation of an Intergovernmental Science-Policy Platform on Biodiversity and Ecosystem Services (IPBES) to facilitate an effective exchange of information and ensuring that policies and actions developed to protect biodiversity are scientifically supported. This will provide Japan with another chance to contribute to global and regional efforts.

Another significant issue remains with recognizing the economic value of biodiversity to promote its integration into public and private decision-making. Recent efforts have involved making biodiversity – its conservation as well as its extinction – of sufficient economic importance for it to be included in decision-making processes by governments and corporations. In this regard, TEEB, a major international initiative, and its reports with compilation of cases around the world including 'forest environmental taxes' in Japan, draw attention to the global economic benefits of biodiversity, serve as a strong mechanism that highlights the growing costs of biodiversity loss and ecosystem degradation, and consolidate expertise from the fields of science, economics and policy to better benefit conservation efforts.

However, parallel to the dimension of governance, is the dimension of security. Unfortunately, it has been demonstrated that climate change has emerged as one of the most important causes of the loss of biodiversity, therefore humanity is now faced with not one, but two major environmental challenges that are closely interconnected. The Millennium Ecosystem Assessment (MEA) has affirmed that climate change is likely to become the dominant direct driver of the loss of biodiversity by the end of the century; the greenhouse gases (GHGs) trapped in the atmosphere and which consequently heat the planet, have already affected diverse habitats and caused the extinction of species like the golden toad and the Monteverde harlequin frog. Conversely, increased protected biodiversity can help reduce the effects of climate change, therefore evidencing the inverse correlation that exists between them.

Biodiversity and Ecosystem Insecurity: A Planet in Peril, precisely reflects on these and other important issues that are interlinked to the state of biodiversity in the planet. The contributions found in this publication address the implications climate change and governance of biodiversity have on security, development and peace, which in turn affect the well-being of humans on Earth. With new inputs and strategies developed in preparation for the high-level meeting on biodiversity – the first in history – during the 65th session of the UN General Assembly, September 2010, this book represents yet another significant contribution to understanding the importance of conserving biodiversity and the implications associated with its loss.

2010 was a milestone year in which the world came together to celebrate life on Earth. The need to halt biodiversity loss has never been as present as it is now and *Biodiversity and Ecosystem Insecurity: A Planet in Peril* provides a comprehensive platform of the challenges we face as a global society and aims to provide input and suggestions on how to address the plethora of issues interacting with a vital source of well-being: biodiversity.

Preface

Ahmed Djoghlaf and Felix Dodds

This book has been produced as a contribution to the discourse on the increasing threats of the biodiversity crisis on international peace and security of the world.

In the 20 years since the beginning of the negotiations for a framework convention on biological diversity, we have seen the loss of biodiversity and ecosystems increase – not diminish – to the point where they are starting to reach tipping points and threatening planetary boundaries. Every living organism on Earth plays an important role in the web of life. The biodiversity present today in our planet is the result of billions of years of natural and human evolution, and it represents an essential resource for human living. Unfortunately, throughout the decades, unsustainable human patterns of consumption of natural resources have led to unprecedented rates of biodiversity loss, species extinction and degraded ecosystems, which have affected not only the habitats of thousands of species, but the livelihoods of many of our own human population.

At the World Summit on Sustainable Development (WSSD), the leaders of the world agreed to substantially reduce the rate of loss of biodiversity by 2010. The third edition of the Global Biodiversity Outlook released early in 2010 by the Convention on Biological Diversity (CBD) demonstrates that this commitment has not been met. It shows that we continue to lose species at up to 1000 times the natural background rate and that ecosystems may be approaching tipping points beyond which widespread and irreversible degradation will take place. It shows that the main drivers of biodiversity loss have not only remained more or less constant over the past decade, but are in some cases intensifying. The report also makes plain that the poor will suffer the most if we cannot reverse these trends. To quote its text:

> *It is clear that continuing with 'business as usual' will jeopardize the future of all human societies, and none more so than the poorest who depend directly on biodiversity for a particularly high proportion of their basic needs.*

The report confirms that we are living in an increasingly unstable world moving towards a planet in peril, a planet in danger of reaching points of no return, where ecosystems and habitats are being lost forever and their capacity to provide resources for future generations is lost along with them. However, amid the grim scenario, there is still belief that there is time to change the course of action to create a more balanced planet for all species on Earth, including us humans. But time is running out, and it is running out fast. As Senator Robert Kennedy said in 1968:

> *A revolution is coming, a revolution which will be peaceful if we are wise enough; compassionate if we care enough; successful if we are fortunate enough – But a revolution is coming whether we will it or not. We can affect its character; we cannot alter its inevitability.*

With the proposal of a new global strategy, the action of world governments in the upcoming years will determine whether the natural habitats and conditions of our planet will be able to continue providing vital resources to present and future generations.

The Challenges

Ecosystem and biodiversity loss which are further compounded by the diverse effects of climate change, pose a great threat to humanity. However, the issues affecting biodiversity and their consequences are multifaceted, therefore we have asked the contributors to expand on what are the current conditions and historical trends of ecosystems and biodiversity; what impacts could these have on increasing insecurity in the world; what have been the consequences of changes in ecosystems for human well-being; what could happen in the future; what should the response be; and what are the decisions that governments singularly or together have to take for, and with, their people to sustain our biodiversity and ecosystems.

This book takes a look into the challenges we as a global society are facing and aims to provide input and suggestions on ways to address such challenges. We have therefore broken down the book into four sections:

Part I: Drivers of Biodiversity and Ecosystem Loss

This section looks at some of the drivers causing biodiversity and ecosystem loss, such as population, poverty, pollution, urbanization and habitat fragmentation, resource exploitation and land use and land-use change.

Part II: Biodiversity Insecurity

This section of the book will present an overview of the potential natural security threats and challenges posed by biodiversity. It will look at the implications biodiversity has on peace and security from environmental, social, economical and political aspects. It will address turning points and resilience, food security, the importance of coral reefs, oceans biodiversity and forest ecosystems.

Part III: Climate Change Biodiversity and Ecosystems

This section will look at the additional impacts on biodiversity and ecosystem loss due to climate change. This will include the vulnerability of states, the role of agriculture, water, mega biodiversity regions, food security, permafrost and tundra loss and the multiple benefits of the United Nations Collaborative Programme on Reducing Emissions from Deforestation and Forest Degradation (REDD) for biodiversity and ecosystem conservation.

Part IV: Governing Biodiversity

Finally, this section of the book will explore the potential avenues and mechanisms available to the international community to address and avert biodiversity and ecosystem loss. The chapters in this section will address a number of forward looking options for international governance, and will focus on the appropriate forums. It highlights the role of economics, stakeholders, the private sector and the importance of funding for biodiversity.

This book is a companion to the *Human and Environmental Security* book from 2005, edited by Felix Dodds and Tim Pippard, and the *Climate Change and Energy Insecurity* book from 2009, which was edited by Felix Dodds, Andrew Higham and Richard Sherman. Both the contributions were nominated for the International Studies Association Environmental book of the year.

Planet in Balance

The first International Year of Biodiversity made it clear to us that the challenges ahead are great. We have the chance in 2012 at the next Rio Earth Summit to change the direction of our economic model to create a balanced planet where resources can be shared and utilized in a sustainable manner, but global action must start now. Our responsibilities include passing on a healthy planet and rich biodiversity and ecosystems to future generations. We can only do that with the support of all stakeholders, including civil society, but most notably policy-making entities that hold within their grasp the course of action that millions can and will take.

In addition, the parallels of ecological problems with the financial crisis are clear. The banks and financial institutions privatized the gains and socialized the losses. We are doing the same with the planet's natural capital. According to the World Wide Fund for Nature (WWF) we are operating at 25 per cent above the biological capacity to support life and that is before adding another billion people by 2020. Though it is hard for some societies to understand and accept that present lifestyles must change, it is absolutely necessary to strive for more sustainable ways of consuming the finite resources on Earth. We are presently drawing upon the ecological capital from future generations and if natural loss continues at this rate, for the first time in history, humanity will face the sixth greatest extinction, the first ever caused by anthropogenic influence, as referenced by the Institute of Sustainable Development and International Relations (2010).

If the Security Council of the UN organization convened, in April 2007, a meeting to discuss the security implication of climate change, time has come also to convene a special meeting of the Security Council on the biodiversity crisis and its security implication. When biodiversity disappears, there is no replacement; when ecosystems fail we cannot recreate them. We need to work together to create a planet in balance where everyone can benefit from living on this planet together, simply because biodiversity is life, biodiversity is OUR life.

Special Thanks

This book would not have been possible without the support of a number of key people to whom we would like to give special thanks: Claire Lamont, Nick Ascroft and Jonathan Sinclair Wilson at Earthscan; Emily Benson, Nicola Williams, Homera Cheema and Robert White at Stakeholder Forum, Prakhar Goel, Natalie Lum-Tai, Philip Otieno and Kiara Worth from the CSD Youth Caucus, Samantha Simic and *a special thanks to* the staff at the UN Secretariat of the CBD who have contributed to this publication.

We dedicate the book to the work of Rachel Carson who inspired us and warned us of the dangers ahead:

> *We stand now where two roads diverge. But unlike the roads in Robert Frost's familiar poem, they are not equally fair. The road we have long been travelling is deceptively easy, a smooth superhighway on which we progress with great speed, but at its end lies disaster. The other fork of the road – the one less travelled by – offers our last, our only chance to reach a destination that assures the preservation of the Earth.*

Contributors

Salvatore Arico is biodiversity and ecosystem services specialist at the United Nations Educational, Scientific and Cultural Organization's (UNESCO) Division of Ecological and Earth Sciences and senior visiting research fellow at the UN University Institute of Advanced Studies in Yokohama, Japan. He is UNESCO's focal point for biodiversity and environmental matters and has been special advisor to UNESCO's assistant director-general for natural sciences (2006–2010). He was also chief of unit, Convention on Biological Diversity (CBD), Montreal, Canada (1998–2000). Among his most significant scientific and policy roles are: invited expert of the Intergovernmental Panel on Climate Change (IPCC) Fifth Assessment Report Scoping Meeting (2009). He has been a contributor author to and reviewer of the Millennium Ecosystem Assessment (MEA) (2001–2005). He holds a PhD in marine environment and resources, Stazione Zoologica 'Anton Dohrn' of Naples, Italy (1992). He has published more than 60 scientific and policy publications.

Monique Barbut, a French national, has been the chief executive officer (CEO) and chair of the Global Environment Facility (GEF) since August 2006. Prior to joining the GEF, Monique served as director of the Division of Technology, Industry and Economics of the United Nations Environment Programme (UNEP). As a member of the French government delegation to the 1992 Rio Earth Summit, Monique was a key player in the financing negotiations, and later on an active negotiator in the creation of the GEF, as well as the French Global Environment Facility, to which she was appointed first CEO.

Craig Bennett is director of policy and campaigns at Friends of the Earth (England, Wales and Northern Ireland), one of the UK's most influential environmental organizations. From 2007 to 2010, he was deputy director at the University of Cambridge Programme for Sustainability Leadership (CPSL) and, in this role, was director of The Prince of Wales's Corporate Leaders Group (CLG) on Climate Change. He is both a senior associate of the Programme for Sustainability Leadership and a visiting programme director at the Judge Business School, University of Cambridge. He is also vice-chair of Stakeholder Forum and a teaching fellow on sustainability at the Department of Engineering at the University of Bristol. Prior to joining CPSL, Craig was the head of the

Corporates and Trade Campaign at Friends of the Earth. He is a regular contributor to National Geographic Green.

Peter Bridgewater is currently chair of the UK Joint Nature Conservation Committee (JNCC), following terms as secretary general of the Ramsar Convention and director of the Division of Ecological Sciences in UNESCO. Prior to those posts he was chief executive of the Australian Nature Conservation Agency from 1990. He has also held many key international appointments in the field of biodiversity, including: member of the board of the MEA (2000–2005); commissioner of the Independent World Commission on the Oceans (1995–1998); and chairman of the International Whaling Commission (1995–1997). He maintains a research interest in international environmental governance, linkages between cultural and biological diversity, and has authored or co-authored more than 200 publications on biodiversity management and conservation, environmental governance and biodiversity issues.

Lim Li Ching has a BSc in ecology and an MPhil in development studies. She is a senior researcher at the Third World Network (TWN), an international non-governmental organization (NGO) based in Malaysia. She coordinates TWN's sustainable agriculture work, which focuses on food security and climate change issues, among others. Lim was a lead author for the East and South Asia and the Pacific sub-global report of the International Assessment of Agricultural Knowledge, Science and Technology for Development (IAASTD). She is also a Senior Fellow at the Oakland Institute, US.

Biliana Cicin-Sain (PhD in political science, UCLA, postdoctoral training, Harvard University) is director of the Gerard J. Mangone Center for Marine Policy and professor of marine policy at the University of Delaware's College of Earth, Ocean, and Environment. Biliana is the organizer, co-chair and head of secretariat of the Global Forum on Oceans, Coasts and Islands, initially mobilized in 2000. Biliana is the author of more than 100 publications in marine policy. Her 1998 book on *Integrated Coastal and Ocean Management: Concepts and Practices* has been used in academic and governmental training efforts around the world. *Integrated National and Regional Ocean Policies: Comparative Practices and Future Prospects* brings together analyses of the experiences of 15 nations and four regions of the world that have taken concrete steps toward cross-cutting integrated oceans governance. In 2007 Dr. Cicin-Sain was awarded the Elisabeth-Mann-Borgese Meerespreis ('Prize of the Sea') and the Coastal Zone Foundation Award for leadership in coastal management and, in 2002, she was awarded the Ocean and Coastal Stewardship Award at the California and the World Ocean Conference together with the late Robert W. Knecht, her husband and co-author, first director of the US Coastal Zone Management Programme.

Peter Costigan is currently the science coordinator for the Environment and Rural Policy Group of Defra in the UK. Peter has a first degree in Biology and a PhD in land reclamation. Following 12 years of research on soil and crop science, Peter joined the UK Ministry of Agriculture in 1991 and managed their agri-environment science for 12 years. In 2003, Peter moved to the Environment and Rural Policy Group in Defra where he oversees evidence on a wide range of environmental issues. He is particularly involved in developing the evidence base to support the ecosystem approach. He is involved in the UK National Ecosystem Assessment, and is chair of the UK Biodiversity Research Advisory Group.

Georgia Destouni is professor of hydrology, hydrogeology and water resources at Stockholm University, and research leader in the Bert Bolin Centre for Climate Research and the strategic research programme EkoKlim (a multi-scale, cross-disciplinary approach to the study of climate change effects on ecosystem services and biodiversity). Her research interests further include hydrological transport of tracers, nutrients and pollutants, and hydro-climatic, groundwater–surface water and freshwater–seawater interactions. Destouni is member of the Royal Swedish Academy of Sciences and its Environmental Committee and Energy Committee, and member of the Royal Swedish Academy of Engineering Sciences.

Ahmed Djoghlaf, an Algerian national, assumed the position of executive secretary of the CBD on 3 January 2006. Previously, he served as director of the UNEP division of the GEF from 1996 to 2005 and was promoted to assistant executive director of UNEP in 2003. In 1994–1995, he was the principal officer in the interim Secretariat of the CBD and assisted in the preparation of the first and second Conference of the Parties. Before joining the UN, Ahmed held a variety of important posts in the Algerian Ministry of Foreign Affairs. He was advisor on environmental issues to the Prime Minister of Algeria, and, prior to that, to three ministers of foreign affairs of Algeria. He holds a PhD from the University of Nancy, France, as well as four other postgraduate degrees. Ahmed holds the rank of minister plenipotentiary of the Ministry of Foreign Affairs and assistant secretary general of the UN.

Felix Dodds is the executive director of the Stakeholder Forum for a Sustainable Future. He has been active at the UN since 1990, attending the World Summits Rio, Habitat II, Rio+5, Beijing+5, Copenhagen+5, World Summit on Sustainable Development (WSSD), World Summit 2005. He has also been to all the UN Commissions for Sustainable Development and UNEP Governing Council. He has set up three global NGO coalitions for UN Conferences, Summits and Commissions: the UN Commission on Sustainable Development (1993), the UN Habitat II (1995), the World Health Organization (WHO) Health and Environment Conference (1999). He co-chaired the NGO Coalition at the UN Commission on Sustainable Development from 1997 to 2001. He introduced stakeholder dialogues through the UN General Assembly in November 1996 for Rio+5 and helped run some of the most successful ones at Bonn Water (2001) and Bonn Energy (2004). He has written or co-edited the following books: *Energy and Climate Change Insecurity: The Challenge for Peace, Security and Development*, *Negotiating and Implementing Multilateral Environmental Agreements*, *Human and Environmental Security: An Agenda for Change*, *How to Lobby at Intergovernmental Meetings: Mine is a Cafe Latte*, *Earth Summit 2002: A New Deal*, *Multi-stakeholder Process on Governance and Sustainability*, *The Way Forward: Beyond Agenda 21* and *Into the Twenty First Century: An Agenda for Political Realignment*. He occasionally writes for the Green Room of the BBC and blogs from film festivals.

Luc Gnacadja is the executive secretary of the United Nations Convention to Combat Desertification (UNCCD). Born in Benin, he is an architect by profession. Before taking up his present position, he served as Minister of Environment, Housing and Urban Development of Benin from 1999 to 2005. He gained firsthand knowledge of the UNCCD process over a number of years in his capacity as head of delegation to the COP to the UNCCD, to the United Nations Framework Convention on Climate Change (UNFCCC) and to the CBD. He has served as chairman of several international ministerial conferences related to the environment and sustainable development. In March 2003, he was honoured with the 2002 Green Award in Washington, DC, by the World Bank.

Gusti Muhammad Hatta is currently the State Minister of Environment of the Republic of Indonesia and has held this post since 2009. With a Bachelor of Forestry in Silviculture from Lambung Mangkurat University, a Masters of Science in Silviculture from Gajah Mada University and a PhD in Silviculture from Wageningen University in The Netherlands, his professional life has included being the vice-rector for academics and head of the Research Institute of Lambung Mangkurat University. His research work has included surveys of flora and fauna of National Forest Kutai, Conservation Area Kayan Mentarang, and Conservation Area Bukit Raya.

Georg A. Heiss is project manager and researcher at the Museum für Naturkunde Berlin and director of Reef Check Germany. He studied geology in Heidelberg, received his PhD at the University Kiel and held postdoc positions at GEOMAR Kiel and at CNRS in Gif-sur-Yvette and Aix-en-Provence. He worked at the German Advisory Council on Global Change, the Centre for Tropical Marine Ecology Bremen and as independent consultant. His main interests are coral reef paleoclimatology, human impacts on coral reefs, reef monitoring and conservation and public outreach programmes.

Jon Hutton is director of the UNEP World Conservation Monitoring Centre where he is responsible for the work of 80 scientists and activities such as the partnership measuring progress towards the global 2010 biodiversity target. Jon has a background in biodiversity conservation and rural development, and worked in Africa for 25 years. He has produced more than 50 papers and book chapters covering issues such as community-based resource management, sustainable use and the relationship between conservation and poverty. Jon is a senior member of Hughes Hall College, Cambridge, and an Honorary Professor of Sustainable Resource Management at the University of Kent.

Yemi Katerere, a Zimbabwean, is a forester by training with experience in forest management and research as well as commercial forestry, including seven years as the chief executive of the Zimbabwe Forestry Commission, seven years heading the International Union for Conservation of Nature (IUCN) regional office for southern Africa and three years as chair of the ICRAF Board of Trustees. He is currently the head of the UN-REDD Programme Secretariat based in Geneva. Yemi has holds a PhD in Forest Resources from the University of Idaho and has published extensively. In recognition of his contribution to forestry and development, Yemi was awarded the Commonwealth Queen's Award in 1993.

Wolfgang Kiessling is professor of paleontology at Humboldt University and a researcher at the Museum für Naturkunde in Berlin. Kiessling received his PhD at the University of Erlangen and held postdoc positions at the University of Erlangen, Humboldt University and the University of Chicago. His research focuses on large-scale analyses of ancient reef systems in terms of ecology and evolution. Wolfgang is especially interested in the bearing of ancient reef crises on current patterns of reef decline.

Johan Kuylenstierna is chief technical advisor to the chair of UN-Water and adjunct professor in International Water Issues at Stockholm University. His work spans the entire water policy field but focuses primarily on global and institutional dimensions of water management and the relationships to climate change, food security, and so on. Johan is a member of the Royal Swedish Academy of Agriculture and Forestry.

Christophe Lefebvre, French scientist, is a geographer graduated at the National Geographic Institute of Paris Sorbonne. He is a former officer at the French coastal conservancy. He has chaired the national council for protection of nature in France for five years. As chairman of the national committee of IUCN and member of the national council of sustainable development, he was one of the main contributors to the national strategy on Biodiversity. He has been an expert of the Ramsar Convention on wetlands of international importance through the Danone Evian programme from 1998 to 2008. Elected as regional IUCN councillor at the last World Conservation Congress in Barcelona, in 2008, he is the voice of ocean advocacy at the IUCN Council. He teaches the law of the marine and costal environment at the University of Côte d'Opale in France and he is a delegate for international affairs at the new French marine protected areas agency.

Jean Lemire, a biologist by training, is an award-winning producer and director of films, such as the popular *Arctic Mission*. With more than 20 years of experience in marine biology, Jean embarked on his greatest adventure on board the *Sedna IV* in 2005. With a team of sailors, film-makers and scientists, he set out to conquer Antarctica, the last continent. Antarctic Mission became one of the greatest expeditions in modern history – 430 days of navigation, isolation and extreme adventure in the unmanageable Antarctic climate. In the autumn of 2010, the International Year of Biodiversity, he embarked on a 1000-day expedition for the planet on board the *Sedna IV*, this time focusing on biological diversity. Jean speaks to raise public awareness of the planet's major environmental issues.

Thomas E. Lovejoy is an innovative and accomplished conservation biologist who coined the term 'biological diversity'. He currently holds the biodiversity chair at the Heinz Center for Science, Economics, and the Environment based in Washington, DC. He served as president of the Heinz Center from 2002 to 2008. Before assuming this position, Lovejoy was the World Bank's chief biodiversity advisor and lead specialist for environment for Latin America and the Caribbean as well as senior advisor to the president of the UN Foundation. Lovejoy has served on science and environmental councils under the Reagan, Bush and Clinton administrations. In 2001, Lovejoy was awarded the prestigious Tyler Prize for Environmental Achievement. In 2009, he was the winner of BBVA Foundation Frontiers of Knowledge Award in the Ecology and Conservation Biology Category. Lovejoy holds BS and PhD (biology) degrees from Yale University.

Ryu Matsumoto has served as the Minister of the Environment and the Minister of State for Disaster Management in Japan since September 2010. He graduated from Chuo University, Faculty of Law in 1970, and was elected member of the House of Representatives for the first time in 1990. He has had a long career at the Japanese parliament, and served as the chairman of the Committee on Environment of House of Representatives in 2002. He was also elected as the chair of the General Assembly of Democratic Party of Japan Diet Members in 2009.

Jan L. McAlpine is director of the UN Division on Forests and head of the United Nations Forum on Forests (UNFF) Secretariat, based in New York. She has served previously as the senior advisor and lead for forests in the US Department of State in Washington, DC. Jan served in the US government from 1989, first with the Environmental Protection Agency and subsequently she worked at the White House, first with the President's Council on Sustainable Development and then in the Office of the US Trade Representative as a negotiator on issues relating to international forest and timber trade. Jan grew up in Francophone Africa, Rwanda, Burundi, Congo, Kenya and South Africa.

Lera Miles has worked on climate change and biodiversity issues since 1997, joining the UNEP World Conservation Monitoring Centre in 2002. Over the past three years, her work has concentrated on the potential impacts of climate change policy on biodiversity and ecosystem services, such as Reducing Emissions from Deforestation and Forest Degradation (REDD), and biofuel development. Previously, she has worked on the regional impact of climate change from the Amazon to the Arctic; and on various scenario exercises, including as part of the GLOBIO3 biodiversity modelling group.

Matea Osti completed a BA in Environmental and Global Studies at Randolph College in Virginia, US, and a Master of Science in Environmental Change and Management at the University of Oxford, UK. She currently works for the Climate Change and Biodiversity Programme at the UNEP World Conservation Monitoring Centre in Cambridge, UK. Matea's main area of work focuses on ecosystem-derived multiple benefits from REDD+. This includes work on carbon, biodiversity and ecosystem service mapping to assist countries with addressing multiple benefits in the planning and implementation of their climate change mitigation measures.

Johan Rockström is Executive Director of Stockholm Resilience Centre, professor in Natural Resources Management at Stockholm University and Executive Director of Stockholm Environment Institute (SEI). He has more than 14 years' experience within water resource management, agricultural development and resilience research. He is also coordinator of several national and regional research and development projects linked to the Global Water Partnership, the Global Dialogue on Water for Food and Environmental Security, and the Resilience Alliance. Johan has also contributed to the management and strategic planning of WaterNet, a regional capacity-building programme in Southern Africa, as well as 40 higher-learning and research institutions in 12 countries.

Linda Rosengren is currently a Natural Resources Officer for the UN-REDD Programme Secretariat, supporting the programme's coordination of global activities related to carbon measurement, reporting and verification, governance, stakeholder engagement and multiple benefits of REDD+. Prior to her work with the UN-REDD Programme, Linda worked as a Forestry Officer for the Food and Agriculture Organization (FAO), where she was in charge of stakeholder consultations supporting the implementation of planted forest best practice guidelines in South-East Asia and South America. She also worked with biofuels, reforestation and community-based forest management technical cooperation projects. Prior joining the UN, Linda worked for the Finnish Ministry of Agriculture and Forestry. She holds a Master in Agroecology.

Christoph Schröter-Schlaack is researcher at the Department of Economics at the Helmholtz Centre for Environmental Research UFZ in Leipzig, Germany. Currently, he works for TEEB scientific coordination and has co-authored different chapters of the TEEB Report for National and International Policy-makers. Before joining TEEB he did research on environmental policy instruments, especially on instruments related to land development and soil protection. Christoph holds a diploma in economics, and is finalizing a PhD in institutional economics.

Maria Schultz is director of SwedBio a knowledge interface contributing to poverty alleviation, sustainable livelihoods, equity and human well-being through development towards resilient ecosystems and societies; funding related initiatives; and facilitating access and participation of developing country stakeholders in related international policy development. Maria had formal education in systems ecology and media communication and her experience ranges from working for an indigenous organization in the Amazon region, to the Swedish Ministry of Environment as CBD focal point, the Swedish International Development Cooperation Agency, universities, and as consultant for civil society.

John Scott is an indigenous descendant from North Eastern Australia. With a Master of (Indigenous) Legal Studies, he has a significant background in education, social policy, law, indigenous rights and traditional knowledge. His professional life has included teacher, Aboriginal education advisor, CEO for Aboriginal programmes, deputy director for the School of Indigenous Studies at James Cook University, manager of the Cultural Unit with National Aboriginal and Torres Strait Islander Commission, indigenous human rights officer with the UN High Commission on Human Rights and second in charge of the Secretariat of the UN Permanent Forum on Indigenous Issues. John is currently programme officer for traditional knowledge for the CBD.

Richard Sherman works as a consultant to the Stakeholder Forum for a Sustainable Future and the South African Department of Environmental Affairs and Tourism with a particular focus on institutional reform and global governance. He was the programme manager for the International Institute for Sustainable Development Reporting Services Africa Regional Coverage Project. He was a former team leader and writer from 2002–2005. In addition, from 1998–2001, Richard was a member of the South African government's climate change delegation, and has played an active role in the South African NGO sector since 1995. His previous books are: *Energy and Climate Change Insecurity*, which he edited with Felix Dodds and Andrew Higham, *10 Days in Johannesburg: A Negotiation of Hope*, with Pam Chasek and Chris Spence; and *Environment and Development Decision Making in Africa 2006–2008*.

Previously he has worked for Global Legislators Organization for a Better Environment (GLOBE) and Earthlife Africa. He has also been an active member of the Climate Action Network.

Nella A. Shpolianskaya graduated from the Faculty of Geography at Moscow State University in 1954. She was working at the Institute of Permafrost Studies, named after V. F. Obruchev, until 1957, when she joined the Department of Cryolitology and Glaciology, in Moscow State University. Currently she is the leading scientist in this department. Nella defended her PhD dissertation in 1965 and her doctoral dissertation in 1979. Key areas of scientific expertise: temperature of rocks as a basis for spatial patterns in cryolitozone in the past, present day and in the future; underground ice in cyolitozone within the system 'shelf/coast' and its palaeogeography. Nella has authored or co-authored 195 publications, including five monographs and two textbooks. She is a member of the Russian Geographical Society, expert of the Scientific-Consultative Centre of the Russian Federation and member of the Doctoral Dissertation Council of the Moscow State University.

Caroline Spelman has been a Member of the UK Parliament since 1997. While in opposition, she served in the Shadow Cabinet, covering the environment, international development and communities and local government portfolios, as well as other senior posts. Before entering parliament, Caroline had an extensive career in the agriculture sector, with 15 years in the agriculture industry and in-depth experience of the international arena, including as deputy director of the International Confederation of European Beet Growers and a research fellow for the Centre for European Agricultural Studies. She has also authored a book on the non-food use of agricultural products.

Achim Steiner serves as the executive director of UNEP and was appointed director-general of the UN Offices at Nairobi in March 2009. Before joining UNEP, Achim served as director-general of the International Union for Conservation of Nature from 2001 to 2006. His professional career has included assignments with governmental, non-governmental and international organizations in different parts of the world. In Washington, he was senior policy advisor of IUCN's Global Policy Unit and, in 1998, he was appointed secretary-general of the World Commission on Dams, based in South Africa. His educational background includes a BA from the University of Oxford as well as an MA from the University of London, with specialization in development economics, regional planning, and international development and environment policy.

Pavan Sukhdev is special advisor to the UNEP's Green Economy Initiative, which includes the Green Economy report, the Green Jobs report and The Economics of Ecosystems and Biodiversity (TEEB) study, of which he is also study leader. As a senior banker, he founded and went on to chair the board of

Deutsche Bank's Global Markets Centre in Mumbai. Beyond his contributions to TEEB and the Green Economy Initiative, his work in this area includes: founding and serving as director of the Green Accounting for Indian States project and serving as president of the Conservation Action Trust.

Marjo Vierros is an adjunct senior fellow at the UN University Institute of Advanced Studies (UNU-IAS). She is also a senior fellow with the UNU-IAS Traditional Knowledge Initiative. She has previously coordinated work related to marine and coastal biodiversity at the Secretariat of the CBD. She was also responsible for the background work undertaken for the previous (2004) in-depth review and update of the CBD programme of work on marine and coastal biological diversity. Marjo has previously worked for research, conservation and UN organizations in several countries. She has extensive experience in tropical marine ecology and coastal management, and has undertaken field research in the Caribbean, Central America, Bermuda and the Pacific. She has also designed and implemented research and educational programmes in Belize, Bermuda, Jamaica, Turks and Caicos Islands and Hawaii. She has authored many publications and was one of the lead authors of the MEA (marine fisheries systems). She has degrees in biology, oceanography and marine biology.

Robert T. Watson's career has evolved from research scientist at the Jet Propulsion Laboratory: California Institute of Technology, to a US Federal Government programme manager/director at the National Aeronautics and Space Administration (NASA), to a scientific/policy advisor in the US Office of Science and Technology Policy, White House, to a scientific advisor, manager and chief scientist at the World Bank, to a chair of Environmental Sciences at the University of East Anglia, the director for Strategic Direction for the Tyndall centre, and chief scientific advisor to the UK Department of Environment, Food and Rural Affairs. In parallel to his formal positions, he has chaired, co-chaired or directed international scientific, technical and economic assessments of stratospheric ozone depletion, biodiversity/ecosystems (the Global Biodiversity Assessment (GBA) and Millennium Ecosystem Assessment (MEA)), climate change (IPCC). During the past 20 years, he has received numerous national and international awards recognizing his contributions to science and the science–policy interface, including, in 2003, the Honorary Companion of the Order of Saint Michael and Saint George from the UK.

Acronyms

AMA	Australian Medical Association
ASSI	Area of Special Scientific Interest
BIMAS	*Bimbingan Massal*
CBD	Convention on Biological Diversity
CCAMLR	Convention on the Conservation of Antarctic Marine Living Resources
CEB	Chief Executives Board
CEO	chief executive officer
CEPF	Critical Ecosystem Partnership Fund
CFC	chlorofluorocarbon
CIA	Central Intelligence Agency
CITES	Convention on International Trade in Endangered Species
CLG	Corporate Leaders Group
CMS	Convention on Migratory Species
CO_2	carbon dioxide
COP	Conference of the Parties
CPSL	Cambridge Programme for Sustainability Leadership
CSD	Commission on Sustainable Development
DLDD	desertification, land degradation and drought
EBSA	European Biosafety Association
EC	European Commission
ECOSOC	Economic and Social Council (UN)
EEZ	Exclusive Economic Zone
EIA	environmental impact assessment
EIT	Economies in Transition
EU	European Union
FAO	Food and Agriculture Organization (UN)

FCPF	Forest Carbon Partnership Facility
FFARF	Federal Forestry Agency of the Russian Federation
FSC	Forest Stewardship Council
GACGC	German Advisory Council on Global Change
GBO	Global Biodiversity Outlook
GCRMN	Global Coral Reef Monitoring Network
GDP	gross domestic product
GEF	The Global Environment Facility – the World Bank, the UN Development Programme and the UN Environment Programme established the multi-billion-dollar GEF in 1990 to fund environmental programmes, especially in the South and EIT.
GEO	Global Environmental Outlook
GHG	greenhouse gas
GLADA	Global Land Degradation Assessment
GLOBE	Global Legislators Organization for a Better Environment
GMEF	Global Ministerial Environment Forum
GOBI	Global Ocean Biodiversity Initiative
GPS	Global Positioning System
Gt CO_2-eq/yr	gigatonnes of carbon dioxide equivalent per year
GPA	Global Plan of Action
ha	hectare
HIICR	Heidelberg Institute for International Conflict Research
IAASTD	International Assessment of Agricultural Knowledge, Science and Technology for Development
IAP	Inter-Academy Panel Statement on Ocean Acidification
ICDP	integrated conservation and development project
IAS	invasive alien species
ICRC	International Committee of the Red Cross
IDB	Inter-American Development Bank
IEG	International Environment Governance
IFAD	International Fund for Agricultural Development
IFF	Intergovernmental Forum on Forests
IFOAM	International Federation of Organic Agricultural Movements
IPBES	Intergovernmental Science-Policy Platform on Biodiversity and Ecosystem Services

IPCC	Intergovernmental Panel on Climate Change
IPF	Intergovernmental Panel on Forests
IPM	integrated pest management
IT PGFR	International Treaty on Plant Genetic Resources for Food and Agriculture
IUCN	International Union for Conservation of Nature
JNCC	Joint Nature Conservation Committee
JPOI	Johannesburg Plan of Implementation
JRC	Joint Research Centre
LBAP	Local Biodiversity Action Partnership
LDC	Least Developed Country (49 countries)
LULUCF	Land-use, Land-use Change and Forestry
MDG	Millennium Development Goal
MEA	Multilateral Environmental Agreement
MEA	Millennium Ecosystem Assessment
MPA	marine protected area
MPCE	monthly per capita expenditure
MSC	Marine Stewardship Council
NASA	National Aeronautics and Space Administration
NDVI	normalized difference vegetation index
NGO	non-governmental organization
NOAA	National Oceanic and Atmospheric Administration
NO_x	Nitrogen oxides
ODA	Oversees Development Assistance
OECD	Organisation for Economic Co-operation and Development
OSPAR	Oslo/Paris Convention for the Protection of the Marine Environment of the North-East Atlantic
PA	protected area
PECS	Programme on Ecosystem Change and Society
PES	payment for ecosystem services
ppm	parts per million
PRSP	Poverty Reduction Strategy Paper
RAF	Royal Air Force
RDA	recommended daily allowance
REDD	Reducing Emissions from Deforestation and Forest Degradation

RSPB	Royal Society for the Protection of Birds
SAC	Special Area of Conservation
SEA	strategic environmental assessment
SEI	Stockholm Environment Institute
SFM	Sustainable Forest Management
SBSTA	Subsidiary Body for Scientific and Technological Advice
SIDS	Small Island Developing States, especially important in relation to the Barbados Plan of Action for SIDS
SO_2	sulphur dioxide
SPA	Special Protected Area
SSSI	Site of Special Scientific Interest
STAR	System for Transparent Allocation of Resources
TEEB	The Economics of Ecosystems and Biodiversity
TFCA	transfrontier conservation area
TWN	Third World Network
UK BRAG	UK Biodiversity Research Advisory Group
UN	United Nations
UNCCD	United Nations Convention to Combat Desertification
UNCLOS	United Nations Convention on the Law of the Sea
UNCED	United Nations Conference on Environment and Development
UNCSD	United Nations Conference on Sustainable Development
UNCTAD	United Nations Conference on Trade and Development
UNDP	United Nations Development Programme
UNDRIP	United Nations Declaration on the Rights of Indigenous Peoples
UNEMG	United Nations Environment Management Group
UNEP	United Nations Environment Programme
UNESCO	United Nations Educational, Scientific and Cultural Organization
UNFCCC	United Nations Framework Convention on Climate Change
UNFF	United Nations Forum on Forests
UNGASS	United Nations General Assembly Special Session
UNOCHA	United Nations Office for the Coordination of Humanitarian Affairs
UNU–IAS	United Nations University Institute of Adanced Studies

WCMC	World Conservation Monitoring Centre
WHC	World Heritage Convention
WHO	World Health Organization
WMO	World Meteorological Organization
WRI	World Resources Institute
WSSD	World Summit on Sustainable Development
WTO	World Trade Organization
WWF	World Wide Fund for Nature

Part I

Drivers of Biodiversity and Ecosystem Loss

1

What Are the Drivers Causing Loss of Biodiversity and Changes in Ecosystem Services?

Robert T. Watson and Peter Costigan

The Conceptual Framework of the Millennium Assessment (MA, 2003) distinguishes between different sets of driving forces that influence changes in ecosystems and their services and therefore also impact on human well-being. These include both direct and indirect drivers, and both exogenous and endogenous drivers. Understanding the different characteristics of these factors is an important step towards devising options for responding to the observed ecosystem change.

Biodiversity and ecosystem services are affected by both direct and indirect drivers. A driver is a natural or human-induced factor that directly or indirectly causes a change in ecosystem services. Direct drivers are those that directly impact on biodiversity and ecosystems, for example, land use conversion, overexploitation, introduction of invasive species, pollution and climate change. Indirect drivers are those that influence the direct drivers of change, for example, economic and population growth resulting in an increased demand for food, fibre, water and energy and agricultural subsidies that promote increased agricultural production. Consequently, it is important to understand the relationship between the indirect and direct drivers of change, which can combine in different ways.

Decision-makers need to be able to identify the drivers that operate in a system in order to understand conditions and trends of ecosystem services, consequences for ecosystems and human well-being, and possible directions these trends might take in the future, and they also need to be able to identify response options. Understanding the factors that cause changes in ecosystems and their services is essential to the design of interventions that enhance positive and minimize negative effects. Each driver has a spatial and temporal scale over which it changes and over which it has an effect on ecosystem services and human well-being. Climate change may operate on the spatial scale of a large region; political change may operate at the scale of a nation or a municipal district. Social-cultural change typically occurs slowly, on a timescale of decades, while economic forces tend to occur more rapidly. Because of the variability in ecosystems, their services and human well-being in space and time, there may be mismatches or lags between the scale of the driver and the scale of its effects on ecosystem services.

Drivers, which are largely scale-dependent, can be controlled to varying degrees by decision-makers. Endogenous drivers are largely under the direct control or influence of a decision-maker. Exogenous drivers are largely beyond the direct control or influence of a decision-maker. The most common local-scale endogenous drivers are land use and land cover change, introduction of new technology and invasive species. Local-scale exogenous drivers are more varied, ranging from natural drivers (such as climate) to economic policy to infrastructure development. Drivers that are exogenous at one particular scale may be endogenous at another (usually coarser) scale. For example, prices for a particular commodity are usually an exogenous factor for a farmer that he or she has little control over, while a national government can influence the prices the farmer receives by regulating the market for this commodity.

Processes and structures can influence the effects of drivers on ecosystem services. These include natural phenomena as well as social, political and economic factors. The effect of an international trade agreement on food availability, for example, might be changed by national-level agricultural policies or the practices of local institutions.

Understanding drivers, their interactions and their consequences for ecosystem services and human well-being is crucial to the design of effective responses. Drivers often operate within sets of other drivers creating interwoven causal processes of ecosystem change. Causal processes of ecosystem change interact with each other, often in synergistic ways. Although many responses target specific problems with ecosystem services, the nature of complex systems means that such responses can have unintended consequences for the multiple interacting drivers that operate in the system and their effects on ecosystem services. Individual drivers may be difficult to affect without impacting on others, and therefore intervening in interactions between drivers is often a more direct way to achieve a desired outcome and enables a more integrated and holistic approach to ecosystem service management.

Actors that effect biodiversity and ecosystem services over a range of spatial scales need to work together, including:

1. Individuals and communities at the field level;
2. Public and private decision-makers at the local and national levels; and
3. Public and private decision-makers at the regional (for example, the European Union (EU)) and international level, through international conventions and multilateral environmental agreements (for example, the Convention on Biological Diversity (CBD), Ramsar, Cites and the United

Nations Framework Convention on Climate Change (UNFCCC)) and international trade agreements (for example, the World Trade Organization (WTO)).

Many of the changes in habitats and their associated ecosystem services around the world are a result of satisfying the increased demand for the provisioning services of food, water, fibre and energy at the expense of biodiversity and regulating, cultural and supporting services. The increase in food production (crops and meat) has outstripped increases in the global population in the past 50 years by converting native forests and grasslands to arable and pastoral land, hence food per capita has increased. However, this increase in food production was associated with major declines in diversity and numbers of plants, terrestrial invertebrates and vertebrates, as well as a wide range of regulating, cultural and supporting ecosystem services. Increases in fishing harvests have resulted in 25 per cent of the oceans being overfished and another 50 per cent were at their maximum sustainable limits.

The key indirect drivers of change, which have resulted – and are projected to continue to result – in significant positive and negative changes to habitats and human well-being throughout the world, are:

- Economic growth: national and per capita income, macro-economic policies, international trade and capital flows. World gross domestic product (GDP) is currently about US$58 trillion with the following distribution: EU 28 per cent, US 25 per cent, Japan 8.8 per cent, China 8.5 per cent, Brazil 2.6 per cent, and India 2.2 per cent. Per capita increases in GDP between now and 2030 are projected to be largest in developing countries, for example, China more than 200 per cent, India around 175 per cent, US/Canada around 50 per cent, Western EU around 60 per cent. These projected increases in per capita GDP should enable a reduction in poverty in Asia, but coupled with an increased demand for water, biological and energy resources, that places increased pressure on all ecosystem services. Changes in world trade policies – for example, agricultural subsidies – will also impact on ecosystem services.
- Demographic changes: population size, age, gender structure and spatial distribution. World population is currently about 6.8 billion people with the following distribution; Asia 60 per cent, Africa 15 per cent, Europe 11 per cent, Latin America 8.5 per cent, North America 5 per cent, and Oceania around 0.5 per cent. The current annual rate of population growth has slowed to about 1.1 per cent, compared to about 2.2 per cent in 1963, but still equates to a population growth of 77 million per year. By 2050, world population is projected to range between 8 and 10.5 billion people, with a best estimate of about 9.2 billion, and the following distribution; Asia 57 per cent, Africa 22 per cent, Europe 7.6 per cent, Latin America 8 per cent, North America 5 per cent, and Oceania around 0.5 per cent. The largest changes in regional distribution are projected to occur in Africa, where a doubling of population is projected in some regions, including the region

that today has the greatest poverty and the greatest reliance on ecosystem services. These changes in population will also be accompanied by increasing life expectancy and an aging population.

- Advances in science and technology: investments in research and development; crop and livestock breeding and species selection; and rates of adoption of new technologies, for example, biotechnology, Global Positioning Systems (GPS), information technology. Advances in genomics should provide the basis for improved crop traits – such as, drought, temperature, salinity, pest resistance and nitrogen-use efficiency – using classical plant breeding and genetic modification. Such advances should assist in reducing the projected loss of agricultural productivity due to human-induced climate change, and meeting the 50–70 per cent projected increased demand between now and 2050, through intensification rather than extensification, thus protecting critical habitats and their ecosystem services. Mechanization of farming practices and drainage technologies; use of agro-chemicals, such as fertilizers and pesticides; mechanization of fishing practices and sonar technology; and energy production have been key drivers of ecosystem change.
- Socio-political: governance and policies (for example, democratization); role of women, civil society organizations and the private sector. Governance and policies, which are sensitive to institutional settings, have significant implications for ecosystem services. For example, in southern Africa a trend towards democratization and increased participation can be observed in some countries, yet non-transparent and corrupt policy systems remain in power in others. Trends towards democratization are often associated with market-oriented economies. An economic regime shift due to changes in political structures can also be observed, for example, when a country joins the EU the introduction of EU policies has often led to major shifts in the agricultural production system. Cooperation rather than competition among actors involved in resource management can lead to improved cooperation for sustainable ecosystem management. Empowerment of women, who play a critical role in agricultural systems in many developing countries yet often have no access to finance, property rights, education and gender-sensitive extension services, could contribute to more sustainable agricultural systems.
- Behaviour change: individual choice (for example, consumption patterns, changing diets) and environmental attitudes. Consumer patterns are changing in two ways: first, consumers in developed countries want to eat fruit and vegetables all year round rather than those that are locally sourced in season, placing pressure on ecosystems around the world; and second, increasing wealth in developing countries is leading to an increase in the demand for meat, which significantly increases the resources needed for agriculture.

These in turn have caused:

- *Conversion and fragmentation of natural habitats:* Human land-use, and agriculture in particular, are important human activities converting natural ecosystems, especially grasslands and forests, resulting in a range of declines in species and populations. This is particularly important when the areas converted are high in species richness or endemism. Globally, the area of cropland is still growing in more than 50 per cent of all countries, although rates of growth are slowing. Meat production is increasing and has grown at an average of about 1–2 per cent per year over the past 50 years. The areal extension of domesticated land (cropland and pasture) over the 20th century ranges from 70 per cent to 80 per cent, and presently increases at a rate of about 0.2 per cent per year. Global round-wood production is the second major field of anthropogenic interference into the world's ecosystems. Forest cover is estimated to have been reduced by about 40 per cent since industrial times and deforestation currently continues at about 15 million hectares annually. Other habitat types, such as tropical, subtropical and temperate grasslands, savannas and shrub-lands have experienced even greater losses.

  Habitat fragmentation can be caused by human-induced land-use changes (for example, clearing natural vegetation for agriculture or road construction) and natural disturbances (such as fires). Small fragments of habitat can only support small populations of rarer species, which therefore tend to become susceptible to extinction. Species that are specialized to particular habitats and those with poor dispersal ability are more adversely affected than generalist species and those with good dispersal ability. About 60 per cent of the world's large riverine ecosystems are highly or moderately fragmented by dams, inter-basin transfers or water withdrawal, resulting in the decline or loss of species and ecosystem services. An increasing pressure that is emerging is the demand for land for production of biomass and biofuels. This is already occupying large areas of land and there is strong pressure to increase the coverage worldwide, in competition with other land uses such as agriculture and natural and semi-natural vegetation. This is likely to have a large effect on habitats.
- *Overexploitation*, especially associated with overfishing in the marine environment, animals hunted for bush-meat, and plants and animals harvested for the medicinal and pet trade: Commercial marine fisheries have in many parts of the world maximized short-term production beyond sustainable levels. This overexploitation has a significant adverse impact in marine ecosystems, both on target species but also on non-target species through wider ecosystem changes. Marine fish populations and communities have changed significantly since the 1960s, with exploited populations declining in abundance. The most vulnerable species are slow-growing and slow-breeding species such as groupers, croakers, sharks and skates, but many prey species are decreasing in average size, as younger cohorts are

exploited. Evidence suggests that many marine populations do not recover from severe depletion even after cessation of fishing. After very marked increases in global fish landings up to about 1990, with a fourfold increase since 1950, landings have been relatively stable since then. The UN's Food and Agriculture Organization (FAO) estimated that, in 2007, about 28 per cent of stocks were suffering from excess fishing pressure and 52 per cent of fisheries were at their maximum sustainable limits. Over exploitation of bush-meat is occurring in many developing countries with tropical forests because of extremely high offtake rates, with gorillas, chimpanzees and elephants being particularly vulnerable.

- *Introduction of exotic invasive species*: Improvements in transportation and globalization have resulted in an increase in both the purposeful (for example, for hunting or biological control) and accidental introduction (e.g. introduced with traded goods or in ballast water) of non-native species. When invasive alien species (IAS) become established – such as in the case of zebra mussels and water hyacinths – they can cause significant ecological, physical and economic damage. IAS can threaten native species as direct predators or competitors, as vectors of disease or by modifying the habitat. IAS have been a major cause of extinctions, especially on islands and in freshwater habitats. While the potentially adverse implications of IAS are well-recognized, the rate of introductions continues to be high and implementation of effective preventive measures are largely lacking. Because of the often non-linear responses to alien species, it can be difficult to quantify the risk they pose for biodiversity. Research has documented about 11,000 alien species in Europe, of which 1094 have documented ecological impacts and 1347 have documented economic impacts (Secretariat of the Convention on Biological Diversity, 2010)
- *Pollution of air, land and water*, especially nitrogen from the use of fertilizers and sulphur from the combustion of fossil fuels: The emissions of sulphur dioxide (SO_2), for example, were responsible for most of the acidification trends observable in the 1970s and 1980s in Europe and North America and now in large regions in Asia. Sulphate aerosols are also major constituents of, in particular, small aerosol particles, having a major climate effect (generally reducing surface temperatures), but also direct impacts on air quality, photosynthetic active radiation, or particle deposition. Estimates of SO_2 emissions since 1850 indicate an almost continuous increase until the early 1990s, with the most pronounced growth occurring since the 1950s (Defra, 2010). Due to increased preventative efforts – such as flue gas desulphurization or switching from high to low-sulphur coal – these trends are slowing down or even reversed. Other relevant emissions by human activities are nitrogen oxides (NO_x) – from fossil fuel combustion and agriculture – and heavy metals, even including toxic substances like mercury. Nitrogen additions to farmland increased significantly in most developed countries between the 1950s and late 1980s. Since then, rates have declined in some developed countries or increased at a slower rate, but nitrogen

consumption in developed countries rose by 6 per cent from 2000 to 2007, and by 32 per cent in developing countries. This produced an overall increase in nitrogen fertilizer consumption of 23 per cent over the seven-year period. There is a similar position with phosphorus fertilizer, where there has been a 2.5 per cent fall in consumption in developed countries from 2000 to 2007, but a 28 per cent increase in developing countries, leading to an 18 per cent increase overall (FAO, 2008). These are very significant additional shifts in nutrient use. Such increases are probably essential to provide the increasing amounts of human food required, but represent a growing perturbation of ecosystems. Developed countries have seen marked deleterious effects of nutrient loadings on the quality of terrestrial, freshwater and marine environments, which has in turn led to controls on use of fertilizers. With the rapid increase now occurring in developing countries we should expect to see similar effects on often vulnerable ecosystems. Concern is also growing regarding the potential impacts of novel pollutants such as endocrine disrupting substances and nano-particles.

There are also direct effects of carbon dioxide quite apart from any effect it has on climate change. The atmospheric concentration of carbon dioxide (CO_2) has increased from its pre-industrial level of about 280 parts per million (ppm) to its current level of nearly 390ppm, and is projected to increase to between 540–940ppm by 2100, without taking into account potential positive or negative feedbacks (Defra, 2010). Increased levels of CO_2 in the atmosphere will change the primary production potential within many habitats. The effect of this can be either damaging or beneficial to different habitats in different situations. There is also growing concern regarding the impact of dissolved CO_2 in the ocean causing ocean acidification. Sea pH levels have fallen from pH 8.2 to 8.1 since the industrial revolution (which, as pH is a logarithmic scale, represents a 30 per cent increase in acidity). The expected biological impact of ocean acidification is uncertain, but many calcifying organisms such as corals, shellfish and some plankton are sensitive to increasing acidity because of reduced availability of calcium ions to make their shells and skeletons.

- *Climate change*: To date this has not been a major driver for most ecosystems, however, it is considered as being a major driver for ecosystem change in the future (MEA, 2005). The Intergovernmental Panel on Climate Change (IPCC) in 2007 reported an observed increase in global mean temperature of about 0.75°C over the 20th century, and stated that it was very likely that most of the observed increase over the past 50 years was very likely (more than 90 per cent certain) associated with human activities. Global mean surface temperatures are projected to increase by 1.4–6.4°C between 2000 and 2100, with land areas warming more than the oceans and high latitudes warming more than the tropics and subtropics. Precipitation is projected to increase at high latitudes and the tropics but decrease in the subtropics, with an increase in heavy precipitation events and a decrease in light precipitation events, hence the risks of more floods and droughts. Arctic sea ice has already declined markedly, with a 30 per cent reduction in the annual minimum area over the past 30 years.

Climate change is projected to affect all aspects of biodiversity – individuals, populations, species distributions and ecosystem composition and function:

1. directly, for example, through increases in temperature, changes in precipitation, loss of ice cover (and in the case of marine systems, changes in sea level, and so on); and
2. indirectly, for example, through climate change altering the intensity and frequency of disturbances such as wildfires.

The most important direct driver during the past 50–100 years was habitat conversion (based on the analysis of the Millennium Ecosystem Assessment (MEA)), especially for forests and grasslands. The next most important were habitat fragmentation, introduction of IAS and overexploitation (especially of marine fisheries). Pollution and climate change have in general had slightly less effect. All direct drivers of biodiversity loss will likely increase on a global scale for the next several decades.

Any 2020 target to halt the global loss of biodiversity will need to take into account a realistic understanding of the multiple indirect and direct drivers of biodiversity and ecosystem services loss, which vary by region, but on average globally are likely to continue to increase because an increasing population and continued economic growth will result in an increased demand for water, biological and energy resources during the next several decades, for example, provisioning ecosystem services. These changes, when coupled with an absence of changes in consumer behaviour and a failure to transition to environmentally friendly technologies, will result in the continued loss of biodiversity and regulating, cultural and supporting ecosystem services. Also:

- The adverse effects of human-induced climate change will continue to increase;
- Deforestation globally will remain a significant challenge and is unlikely to be halted by 2020;
- Regional pollution (air, land and water) in developing countries will likely continue and even increase over the next decade;
- Overexploitation of the oceans will continue as a risk; and
- The introduction of invasive species will be hard to stop.

The importance of climate change and pollution are likely to become dominant drivers of the loss of biodiversity and ecosystem services during the next several decades unless there is a significant reduction in emissions of greenhouse gases (GHGs) – for climate change – and local and regional pollutants. Limiting the impact of human-induced climate change will require equitable long-term coordinated global action to reduce GHGs.

One encouraging factor in assessing the drivers is the great improvement there has been in the collation and availability of the relevant statistics of how they are changing over time. However, there is still a need for more analysis regarding how they impact on biodiversity both singularly and through their interactions, and on the most effective ways of responding in order to protect biodiversity.

References

Defra (2010) 'The Chief Scientific Adviser's report to the Scientific Advisory Committee', SAC (10) 35, Defra, available at http://sac.defra.gov.uk/wp-content/uploads/2010/10/SAC-10-35-CSA-Report.pdf

FAO (2008) www.fao.org

MEA (Millennium Ecosystem Assessment) (2005) *Ecosystems and Human Well-being: Synthesis*, Island Press, Washington, DC

Secretariat of the Convention on Biological Diversity (2010) *Global Biodiversity Outlook 3*, Secretariat of the Convention on Biological Diversity, Montréal, available at www.cbd.int/doc/publications/gbo/gbo3-final-en.pdf

Part II

Biodiversity Insecurity

2

Biodiversity, Peace and Security

Ahmed Djoghlaf

According to the World Meteorological Organization (WMO), the first decade of the new millennium was the warmest on record (WMO, 2009). The year 1998 was recorded to be the warmest year since records began. However, 2010 figures released by the National Oceanic and Atmospheric Administration (NOAA) suggested that 2010 was on course to be the warmest year since records began in 1880, with large parts of Canada, Africa, Europe and the Middle East facing extreme heatwaves and abnormally warm temperatures (NOAA, 2010). In what seems to be a particularly paradoxical life event, 2010, the year declared by the United Nations (UN) as the International Year of Biodiversity, will also be remembered as one of the years in which millions of people, in almost all continents, suffered their worst environmental disaster. The year began with Colombia declaring a state of environmental emergency in 25 of its 32 provinces so as to allow the purchase of firefighting equipment without going through procurement bidding procedure. The year also began with Haiti's devastating earthquake, which killed more than 200,000 people and left almost 2 million homeless. It was soon followed by the 8.8 magnitude earthquake in Chile, which left mass devastation. Russia, witnessing its warmest summer in more than 1000 years, suffered one of its worst environmental disasters in decades. More than 800,000 hectares of land were destroyed by fire leading to the Prime Minister's ban on all export of grain, which consequently lead to a 70 per cent price increase of wheat within one month of the ban. In the memory of Tokyo and Osaka citizens, August was the warmest month. Pakistan experienced its worst flooding disaster in more than 80 years. With almost one fifth of the country underwater, nearly 20 million people were displaced, 10 million left homeless and 1600 lost their lives. Heavy rains in the southern part of Mexico affected more than 100,000 people and Guatemala declared a state of emergency when days of heavy rain caused widespread flooding and landslides. Southwestern China suffered the worst drought in a century and 51 million people were faced with water shortages; however, heavy rains and floods have also affected 18.3 million people in the country. Sanya, in the Hainan province of China, was hit by Mindulle, the fifth tropical storm of the year, while the Gensu mudslide was the deadliest individual disaster among the 2010 China floods. At the same time, half of Niger's population is suffering from famine and malnutrition owing to severe drought. During the 20th century, North Kenya suffered 28 episodes of severe drought, with 4 during the last decade. According to the UN Office for the Coordination of Humanitarian Affairs (UNOCHA), at least 400 pastoralists have been killed in armed conflicts across Kenya since July 2008 (UNOCHA, 2008). The worst hit districts of Turkana, Marsabit and Mandera recorded more than 50 deaths each.

In the Americas, the Deepwater Horizon oil spill in the Gulf of Mexico has had a severe environmental impact on the marine and wildlife habitats of the region, affecting not only thousands of species and animals, but thousands of human livelihoods as well. The explosion of the platform on 20 April 2010 led to almost 5 million barrels of oil spewing into this unique marine ecosystem during 90 days. In an interview with US political website Politico, President Obama compared the Gulf oil spill catastrophe to the terrorist attacks on 11 September 2001, saying: 'In the same way that our view of our vulnerabilities and our foreign policy was shaped profoundly by 9/11, I think this disaster is going to shape how we think about the environment and energy for many years to come' (Politico, 2010). His statement is not the first to link environmental threat to security issues. A report released in 2007 by CNA Corporation, concluded that global climate change presents a serious national security threat that could affect US citizens, impact its military operations and heighten global tensions. The report explores ways in which climate change poses as a threat multiplier in already fragile regions of the world, exacerbating conditions that provide breeding grounds for extremism and terrorism (CNA Corporation, 2007).

A similar conclusion was reached by the report commissioned in 2008 by former Prime Minister of the UK, Gordon Brown, to outline the new defence strategy, stating that 'climate change is potentially the greatest challenge to global stability and security' (BBC, 2008). Addressing the House of Commons, Prime Minister Brown stated: 'The nature of the threats and the risks we face have – in recent decades – changed beyond recognition and confound all the old assumptions about national defence and international security' (BBC, 2008); he also added that climate change and pandemic disease threaten international security as much as terrorism and that Britain must radically improve its defences. In addition, Sir David King, science advisor to former British Prime Minister, Tony Blair, stated that 'climate change is a far greater threat to the world than international terrorism' (in Khor, 2006).

Following the meeting of the Security Council of the UN devoted to climate change, energy and security, the former UN Secretary-General, Kofi Annan, stated that the biggest threat to humanity is climate change. As early as 1988, the Prime Minister of the UK, Margaret Thatcher, requested a special meeting of the UN Security Council to address the threats of climate change. The meeting, that took place on 17 April 2007, convened a day-long Security

Council debate on the impact of climate change on security and featured interventions by more than 50 speakers. In the opening remarks, Margaret Beckett expressed the importance to tackle climate change as part of security issues, and added 'this is a groundbreaking day in the history of the Security Council, the first time ever that we will debate climate change as a matter of international peace and security' (UN, 2007). Two months later, the UN unanimously adopted a resolution titled 'Climate change and its possible security implications'. The same year, the Nobel Prize was awarded to the Intergovernmental Panel on Climate Change (IPCC) and former US Vice-President Al Gore. Wangari Maathai was the first environmental activist to receive the Nobel Peace Prize. In announcing the decision of the chair of the Nobel Peace Committee, Ole Danbolt Mjøs, stated, 'environment protection has become yet another path to peace' (Mjøs, 2004).

UN Secretary-General, Ban Ki-moon, has said that 'because the environment and natural resources are crucial for building and consolidating peace, it is urgent that their protection in times of armed conflict be strengthened. There can be no durable peace if the natural resources that sustain livelihoods are damaged or destroyed' (UNIS, 2009). Indeed the gravity of the impact of human-induced destruction of the natural capital of our planet has reached such a level that it poses a major threat to peace and security in the world. The third edition of the United Nations Convention on Biological Diversity's (CBD) Global Biodiversity Outlook (Secretariat of the CBD, 2010), shows that today we humans continue to drive species extinct at an unprecedented rate. This comprehensive report on the status of biodiversity in 2010, based on 120 national reports submitted by parties, demonstrates that that biodiversity continues to disappear at an unprecedented rate – up to 1000 times the natural background rate of extinction. The report further warns that irreparable degradation may take place if ecosystems are pushed beyond certain tipping points, leading to the widespread and irreversible loss of biological goods and ecosystem services that we depend on for our health and well-being. The supremacy of human beings on other living species, though uncontrolled, is not a triumph. The human being is now jeopardizing life on Earth as well as his own existence.

Since the industrial revolution, humanity has entered an era of global changes that are transforming the face of the Earth and its effects will ripple down for centuries. Living standards are generally, increasingly, persistently tapping on natural resources. As such, humanity's footprint has never been as significant. The pressures on the planet's natural functions caused by human activity have reached such levels that ecosystems' ability to satisfy the needs of future generations is seriously, and perhaps irretrievably, compromised. Humanity is facing two main global challenges that are interconnected: climate change and biodiversity loss. These challenges have further implications for peace, security and development.

There is an intricate connection between the state of the environment and biodiversity, and the state of peace or conflict among regions across the world.

Biodiversity helps maintain equilibrium in many environments, such as marine and forest ecosystems. Such equilibrium is necessary to maintain the strength of ecosystems to deal with natural disasters or issues such as pollution and climate change. An ecosystem that is not diverse both genetically and in terms of species cannot withstand long-term stresses without beginning to disintegrate. These ecosystems not only provide habitat for important animal and plants species, but also provide a variety of ecosystem services that benefit humans. Human populations are dependent on these services for their livelihood and the current decline in biodiversity worldwide is jeopardizing many areas of life. For example, most medicinal remedies, even when chemically processed, derive from plants found in diverse ecosystems. These plant species are crucial to the continued health of human populations and their availability is dependent on our sustainable use of resources, as well as our respect for biodiversity. Furthermore, economic security is being threatened by biodiversity loss, for example, due to the billions of dollars spent on natural disaster reconstruction. Another important factor of biodiversity is the social and cultural importance of biodiversity to human beings. Traditional and indigenous communities not only depend on biodiversity for their livelihoods, they consider ecosystems as part of their heritage and they tend to have a spiritual relationship with them.

Degrading Ecosystems and Limited Resources: Implications for Peace

Environmental degradation, either caused by the multiple effects of climate change, or human activity (including deforestation for logging), urban development or resource exploitation, has been linked to political unrest that has put social pressure on vulnerable populations.

A vast majority now agrees that anthropogenic activities are affecting the planet's climate. In turn, climate change is impacting biodiversity and people around the world. Observational evidence from all continents and most oceans shows that many natural systems are being affected by climate change, particularly temperature increases. Recent warming is strongly affecting biological systems, including changes such as earlier timing of spring events, leaf-unfolding, bird migration and egg-laying and poleward and upward shifts in ranges in plant and animal species. The fourth assessment report of the IPCC (2007) highlights a number of regional impacts that could have severe repercussions on the environment and people. For example, by 2020, between 75 million and 250 million people in Africa are projected to be exposed to increased water stress due to climate change. By 2050, climate change will lead to gradual replacement of tropical forest by savanna in eastern Amazonia. Semi-arid vegetation will tend to be replaced by arid-land vegetation. There is a risk of significant biodiversity loss in many areas of tropical Latin America. In North America, coastal communities and habitats will be increasingly stressed by climate change impacts interacting with development and pollution.

Issues such as sea-level rise and unsustainable human development are contributing to the loss of coastal wetlands and mangroves and increased damage from coastal flooding. Global temperature increases of 3–4°C could result in 330 million people being permanently or temporarily displaced through flooding, including some 70 million people in Bangladesh, 22 million in Vietnam and 6 million in Lower Egypt among those affected (UNDP, 2007). This migration often results in additional biodiversity degradation in the areas of relocation, due to an increased population and resource demand. Such migration and environmental pressures are linked to both human health and economical insecurities, affecting people's livelihoods and stability. These risks can also jeopardize fragile peace and development states in post-conflict societies. In Afghanistan, warfare and institutional disintegration have combined to take a major toll. In a clear case of environmentally induced displacement, tens of thousands of people have been forced from rural to urban areas in search of food and employment. In the coming decades, displacement issues due to rising sea levels, even in urban centres, will continue to be an increasing security problem. According to IPCC, impacted areas will reach 60 per cent of the largest urban zones of the world, with a total population of more than 5 million, as well as 12 major cities with populations of more than 10 million (IPCC, 2007). The result of changes in the ecosystems of those areas, including reduced access to land and living space, will most definitely affect both local and international communities.

Climate change certainly exacerbates the situation of water shortage experienced by many countries in a manner that has never been seen since man appeared on Earth. Today, four out of ten people in the world live in countries with a severe shortage of drinking water. The UN estimates that by 2025, 48 nations, with combined a population of 2.8 billion, will face freshwater 'stress' or 'scarcity' (Water.org, undated). The consumption of drinking water today is estimated at 135 litres per person per day in developed countries, but at only 14 litres per person in Africa. Today, in 2010, nearly 1 billion people lack access to safe water and 2.5 billion do not have improved sanitation, resulting in staggering health and economic impacts. Daily, 6000 people, mainly children, die as a result of poor or non-existent sanitation or for want of clean water. According to the 2006 UN *Human Development Report*: 'The water and sanitation crisis claims more lives through disease than any war claims through guns' (UNDP, 2006). According to the Australian Medical Association (AMA), climate change could cause a significant increase in diseases across Asia and Pacific region leading to conflicts and leaving millions displaced. The AMA predicts that, in 2100, more than 15,000 people in Australia will die each year owing to heatwave stress. Through geographic changes in weather patterns, rainfall and temperature, climate change is predicted to increase dramatically the extent and prevalence of some vector borne diseases such as malaria and dengue fever. Extreme weather events may also increase vulnerability to water, food or person-to-person borne diseases such as cholera and dysentery, and lead to increases in heat-related mortality and illness.

According to a Central Intelligence Agency (CIA) report the shortage of water will, in the near future, constitute one of the major sources of tension and armed conflict in the world. According to this report, more than 30 countries receive more than a third of their consumption of drinking water from outside their borders. Such a forecast takes on its true meaning when we consider the fact that, out of the 268 international river basins shared by 145 countries and feeding 40 per cent of the world's population, more than 158 are not governed by any form of joint cooperation mechanism between neighbouring countries. In some cases more than 16 countries share this natural resource. This is the case, for example, of the river Congo, the Niger River and even the Nile. Tension usually builds over access to water sources and its management. Such is the example of the 'Cochabamba Water Wars' in 2000, where thousands of Bolivians protested the privatization of the municipal water supply; or the case of the village of Rabdore in Somalia, where a battle erupted between two clans over control of a water hole, which ended in two years of conflict and more than 250 dead (CRS, 2009). Historian Herodotus called Egypt 'the gift of the Nile', it is therefore not surprising that the signature on 14 May 2010 of the Framework Cooperation Agreement by Tanzania, Uganda, Rwanda and Ethiopia, countries establishing a permanent commission to manage the River Nile's water, created unprecedented tension with Egypt and Sudan. The agreement was considered as the death sentence of Egypt; Russian media reported that the state-owned arms export company Rosoboronexport signed with Uganda an agreement for six sophisticated Russian-made Surkoi Su 3 fighter jets worth US$1.2 billion. The report was denied and Uganda sent its six MIG 21 fighters for overhaul to Russia. In southern Africa, Zimbabwe, Botswana, Tanzania, Mozambique, Namibia and Angola are having a hard time trying to convince Zambia to ratify the Zambezi water use agreement.

The Secretary-General of the UN, Ban Ki-moon, warned that 'throughout the world, water resources continue to be spoiled, wasted and degraded. The consequences for humanity are grave. Water scarcity threatens economic and social gains and is a potent fuel for wars and conflict' (Lewis, 2007). Indeed, the experts from the UK issued an International Alert Report identifying 46 countries – home to 2.7 billion people – where climate change and water related crises will create a high risk of violent conflict. A further 56 countries, representing another 1.2 billion people, are at high risk of political instability.

According to the report by IPCC (2007), global warming and its corollary action of melting snow and glaciers caused, in the 20th century, a 10–20cm increase in sea level. There is the danger, therefore, that global warming will jeopardize the existence of 160,000 glaciers in the world. In September 2003, the largest glacier, Ward Hunt, located in the Arctic, that had existed for more than 3000 years, shattered. This same report stated that global warming could lead, towards the end of the present century, to an increase in the level of the seas of approximately 88cm. This will have disastrous consequences on the world population, as more than 50 per cent of the major cities of the world are located in coastal areas, with more than 3 billion people living less than 100km from the coast. As an example, a study by oceanographer Sugata Hazra found that in the past 30 years, nearly 31 square miles of the Sundarbans Islands have vanished entirely (Sengupta, 2007). The very existence of several island countries such as Tuvalu or Kiribati is under threat. In November 2001, after having been refused by Australia, the authorities of Tuvalu requested New Zealand to provide shelter to 11,000 of its citizens out of fear that the country would soon be submerged. Tuvalu will be recalled as the first country to endeavour evacuating its population because of rising waters, but it will not be the last. Around 65 per cent of the Maldives land mass is barely 1m above sea level. In a statement delivered in October 1987, before the UN General Assembly, the President of the Maldives stated that his country was under threat from the rising sea waters. He described his country of 311,000 people as a 'nation in danger' (in Rebello, 2009). To draw the attention of the world to the security dimension of climate change on the survival of small island states, the President of Maldives, Mohamed Nasheed, convened on 21 September 2009, on the island of Girifushi, an underwater cabinet meeting. Addressing the UN General Assembly, on 19 October 2009, he stated that climate change is equal to a mass murder. The Prime Minister of Grenada, Tillman Thomas, at the Oasis Summit held on 21 November 2009, stated that the incapacity to act on climate change is a silent genocide.

Thus, for the first time in the history of mankind, the rise in sea level is threatening to cause the disappearance of a number of sovereign nations from the surface of the Earth, which no army in the world, however sophisticated, can stop. It is therefore not surprising that a study commissioned by the US Department of Defense concluded that climate change would change US national security in the way that it should be considered immediately (Schwartz and Randall, 2003). According to this study from the Pentagon, the plausible consequences include famine in Europe and possible showdowns over who controls what is left of the world's water.

Depleted water sources and overused infertile lands have led to food and water shortages, leading to intra-state conflicts as people become desperate to provide for their families and sustain their livelihoods. In fact, according to Raleigh and Urdal (2007), environmental degradation causes more internal conflict than international conflict. This is ever more present in developing countries, where a large part of the rural population relies on natural resources. The consequences of precarious conditions for food production from devastated ecosystems – caused either by climate change, natural disasters or anthropogenic influence – are dramatic. Infrastructure is destroyed, disease spreads faster, arable lands are lost and people can no longer grow crops or raise livestock, all directly impacting the health of vulnerable populations. The number of hungry people in the world has passed the 1 billion mark for the first time ever (STWR, 2009), which translates into one in every six people struggling with hunger on a daily basis. For instance, drought-affected areas in sub-Saharan Africa could expand by 60–90 million hectares, with dry land zones suffering losses of US$26 billion by 2060 (UNDP, 2007). It is expected that in Latin America and Asia, agricultural production will be lost and therefore rural poverty is bound to increase and cause conflict. In order to feed the increasing world population, food production must double by 2050. This is a major challenge considering the continuous degradation of soils, desertification and rising sea levels that cause land immersion and population displacement. These stressors often lead to social and economical insecurity, which can aggravate social tension and lead to violent conflict. For example, in many countries in Latin America, environmental degradation, along with exclusionary agrarian reforms and systems, have led to the squatting of thousands of peasants in private lands. These so-called 'invasions' have been the source of violent confrontations between private landowners, military and squatters.

For many countries, the abundance of resources such as oil or water can provide both regional and international power. Any threat to the supply of this resource is likely to justify military action in the name of economic preservation and national security. Thus, international conflict can arise when either economical or social factors interact with resource availability. Since biodiversity degradation caused by factors such as deforestation, overfishing, soil erosion and water shortages are major causes for population relocation and loss of security, tensions can rapidly escalate between nations that share these resources. Ultimately, both abundance (diamonds, minerals, oil, timber) as well as a scarcity of marketable resources contribute to the aggravation of conflict between groups. The former often causes illegal, unsustainable and forceful exploitation of the resource. For example, the illegal poaching and trade of gorilla meat in the Democratic Republic of the Congo has been perpetuated by militia forces in order to fund their violent activities. This factor, as well as the continual degradation of gorilla habitat due to infrastructure growth, is resulting in the endangerment of this species. Conflict also arises when local and indigenous communities attempt to fight for the preservation of the biodiversity upon which they rely to sustain themselves economically and culturally, whether with government factions or corporations. For example, the Penan people of Malaysian Borneo fought against industrial loggers entering their ancestral home. Unfortunately, they were not able to preserve either the forest or their spiritual sites.

Ecocide and Consequences of War

A second perspective when discussing the relationship between peace and biodiversity is the effects war has on the environment and loss of species and habitats. Throughout time, wars around the globe have invariably resulted in the devastation of forests and biodiversity. Modern military technology has resulted in severe widespread environmental impacts, making the term 'ecocide' a fitting one. Water contamination and destruction of water sources, hindering of conservation efforts and depletion of intact forests, are only a few of the multiple environmental consequences of war. Existing intra-state and interstate armed conflicts have direct, and often long-lasting, effects on biodiversity in conflict areas. For example, dense populations in the Gaza refugee camps lead to aquifer depletion and effectively caused saltwater to contaminate water sources, making it unsuitable for agricultural irrigation. Landmines left behind in post-conflict zones bar people from productive lands and force them to encroach on other biodiversity-rich areas in order to sustain themselves by clearing forests for agriculture. For example, the civil war in Sri Lanka led to the cutting of about 5 million trees, robbing many poor people of medicinal plants and food sources found in the forests (UNEP, undated). The abuse of biodiversity by armed groups is also due to other reasons, such as their subsistence while occupying a territory. Armed groups may deforest a small or large area to facilitate access to particular parts of the forest or to reduce the risks of ambush, even if it happens to be a protected area of national park, as has been seen to happen in the Rwandan Virunga National Park. The Darién Province in the south of Panama, which is a dense tropical forest, has suffered a great impact due to the Colombian civil war and the penetration of guerillas from the Revolutionary Armed Forces of Columbia in this area. As a result, the region is now recognized as one of the most dangerous places in the world, where conservation efforts are hindered by violence and where excessive deforestation, poaching and overuse of land threatens the fragile ecosystems of the Darién Gap (Trab Nielsen, 2006).

In addition, environmental degradation and the loss of biodiversity is not always a casualty of war. Harmful environmental devastation also occurs through deliberate acts of military sabotage, such as the torching of oil fields, the destruction or contamination of a water source and the widespread use of defoliants. For example, in 1943, the British Royal Air Force (RAF) bombed dams on the Möhne, Sorpe and Eder Rivers in Germany. The resulting Möhne Dam breech killed more than 1200 people and destroyed all downstream dams for 50km (CRS, 2009). In 2003, a sabotage bombing of a 1.8m-diameter water supply pipeline in Baghdad, Iraq, took place (CRS, 2009).

There is increasing awareness that national security and ecological conservation are closely linked. The impact of war and armed conflicts on the environment, and more specifically on biodiversity, is an important and recurring concern for the international community. According to the Heidelberg Institute for International Conflict Research (HIICR), conflicts worldwide have increased from 272 in 1997, to 365 in 2009 (HIICR, 2009). Of those 365, 80 were conflicts directly over resources, ranking second after ideological conflicts. More than 50 per cent of these resource conflicts were ranked either as violent or highly violent. These conflicts affect millions of people and must be addressed at the international level. The United Nations Environment Programme (UNEP) reports that more than 40 per cent of intra-state conflicts are linked to natural resources (UNEP, 2009) As a response to environmental conflicts, the international community has established several legal agreements and laws aimed at protecting victims of conflicts and preventing environmental damage. For example, an additional protocol to the 1949 Geneva Convention prohibited, in 1977, the use of war tactics that could harm the environment in a serious and/or long-term fashion (ICRC, 2007). Other instruments include the Convention on the Prohibition of the Use, Stockpiling, Production and Transfer of Anti-Personnel Mines and on Their Destruction (1997) as well as the International Committee of the Red Cross (ICRC) Guidelines for Military Manuals and Instructions on the Protection of the Environment in Times of Armed Conflict (1996). These regulations help the international community both prevent and respond to environmental conflict; however, they can only be effective if they are implemented. The implementation and enforcement of these instruments remains very weak. There are few international mechanisms to monitor infringements or address claims for environmental damage sustained during warfare. In 2001, the UN established the International Day for Preventing the Exploitation of the Environment in War and Armed Conflict in an effort to increase knowledge and understanding concerning environmental and biodiversity impacts of war (UN, 2001). This day focuses on encouraging member states to expand international law on environmental protection in times of war and adapt their responses to reflect the predominantly internal nature of today's armed conflicts.

It is very important to include biodiversity protection plans when peace-building, as they contribute to economic development and enhance community and government cooperation. Aside from legal frameworks at the international level, other initiatives have been created in an attempt to harmonize humans and the environment. For example, peace parks, also known as transfrontier conservation areas (TFCAs) have successfully ensured coexistence between humans and nature in Africa, promoting peace and stability and conserving biodiversity, and consist of governments jointly managing natural resources across political boundaries (Peace Parks Foundation, undated). But much more needs to be done. The multiple challenges that environmental destruction has on peace, security and development affect the well-being of all humans on Earth, now and for generations to come. What remains clear is that war and conflict are completly man-made affairs and we, as a global society, must act now to ensure balance between political affairs and the natural environment.

If, in 2007, the UN Security Council had a meeting on the security dimension of climate change (UN, 2007), the time has now come to convene a meeting on the security dimension of the unprecedented loss of biodiversity compounded by climate change. As Martin Luther King Jr said:

> *Human progress is neither automatic nor inevitable. We are faced now with the fact that tomorrow is today. We are confronted with the fierce urgency of now. In this unfolding conundrum of life and history there is such a thing as being too late ... We may cry out desperately for time to pause in her passage, but time is deaf to every plea and rushes on. Over the bleached bones and jumbled residues of numerous civilizations are written the pathetic words: Too late.* (King Jr, 1967)

References

BBC News (2008) 'In full: Brown security statement', http://news.bbc.co.uk/2/hi/uk_news/politics/7304999.stm

Central Intelligence Agency (2000) 'Global patterns' report, the U.S. Central Intelligence Agency (CIA), Langley, VA

CNA Corporation (2007) 'National security and the threat of climate change', http://securityandclimate.cna.org/report/National%20Security%20and%20the%20Threat%20of%20Climate%20Change.pdf

CRS (Catholic Relief Services) (2009) *Water and Conflict*, Catholic Relief Services, Baltimore, MD

HIICR (Heidelberg Institute for International Conflict Research) (2009) *Conflict Barometer*, Heidelberg Institute for International Conflict Research, Heidelberg

ICRC (International Committee of the Red Cross) (2007) 'Factsheet on the 1977 protocols additional to the Geneva Conventions', www.icrc.org/web/eng/siteeng0.nsf/html/protocols-1977-factsheet-080607

IPCC (Intergovernmental Panel on Climate Change) (2007) 'The ARA4 synthesis report', *Fourth Assessment Report: Climate Change 2007*, Intergovernmental Panel on Climate Change, Geneva

Khor, M. (2006) 'Global trends', in *Selected Articles by Martin Khor on Climate Change 2005–2007*, available at www.twnside.org.sg/title2/climate/climate.change.doc, p10

King, M. L. Jr (1967) *Where Do We Go from Here: Chaos or Community?*, Harper & Row, New York

Lewis, L. (2007) 'Water shortages are likely to be trigger for wars, says UN chief Ban Ki Moon', *The Times*, 4 December, www.timesonline.co.uk/tol/news/world/asia/article2994650.ece

Mjøs, O. D. (2004) 'The Nobel Peace Prize 2004', www.greenbeltmovement.org/a.php?id=35

NOAA (National Oceanic and Atmospheric Administration) (2010) 'State of the climate global analysis', www.ncdc.noaa.gov/sotc/?report=global

Peace Parks Foundation (undated) 'What are peace parks/TFCAs?', www.peaceparks.org/Content_1020000000_Peace+Parks.htm

Politico (2010) 'Obama: Gulf spill "echoes 911"', www.politico.com/news/stories/0610/38168.html

Raleigh, C. and Urdal, H. (2007) 'Climate change, environmental degradation and armed conflict', *Political Geography*, vol 26, no 6, pp627–736

Rebello, L. (2009) *World Without Wars*, www.recim.org/stud/Rebello-wwwars2.pdf

Schwartz, P. and Randall, D. (2003) 'An abrupt climate change scenario and its implications for United States national security', www.climate.org/topics/PDF/clim_change_scenario.pdf

Secretariat of the CBD (Convention on Biological Diversity) (2010) *Global Biodiversity Outlook 3*, Convention on Biological Diversity, Montreal

Sengupta, S. (2007) 'Sea's rise in India buries islands and a way of life', *The New York Times*, 11 April, www.nytimes.com/2007/04/11/world/asia/11india.html

STWR (Share the World's Resources) (2009) 'Number of chronically hungry tops one billion', www.stwr.org/food-security-agriculture/number-of-chronically-hungry-tops-one-billion.html

Trab Nielsen, S. (2006) 'ICE case study: The spillover effect of the Colombian conflict ecological damage in the Darién Gap', www1.american.edu/ted/ice/darien.htm

UN (United Nations) (2001) 'International Day for Preventing the Exploitation of the

Environment in War and Armed Conflict', www.un.org/depts/dhl/environment_war/index.html
UN (2007) 'Security Council hold first ever debate on impact of climate change on peace, security, hearing over 50 speakers', www.un.org/News/Press/docs/2007/sc9000.doc.htm
UNDP (United Nations Development Programme) (2006) *Human Development Report 2006*, Palgrave Macmillan, New York
UNDP (2007) *Human Development Report 2007/2008*, Palgrave Macmillan, New York
UNEP (United Nations Environmental Programme) (undated) 'In defence of the environment, putting poverty to the sword', www.unep.org/Documents.Multilingual/Default.asp?ArticleID=3810&DocumentID=288
UNEP (2009) *From Conflict to Peacebuilding: The Role of Natural Resources and the Environment*, UNEP, Nairobi, available at www.unep.org/pdf/pcdmb_policy_01.pdf
UNIS (United Nations Information Service) (2009) '"There can be no durable peace if the natural resources are damaged or destroyed": Message on the International Day for Preventing the Exploitation of the Environment In War and Armed Conflict, 6 November 2009', press release, www.unis.unvienna.org/unis/pressrels/2009/unissgsm149.html
UNOCHA (United Nations Office for the Coordination of Humanitarian Affairs) (2008) 'Horn of Africa crisis report', United Nations Office for the Coordination of Humanitarian Affairs, Nairobi, Kenya.
Water.org (undated) 'Global water supply high school curriculum', http://static.water.org/docs/curriculums/WaterOrg%20HighCurricFULL.pdf
WMO (World Meteorological Organization) (2009) 'WMO statement on the status of global climate change in 2009', www.wmo.int/pages/prog/wcp/wcdmp/documents/WMOStatement2009.pdf
World Climate Report (2005) 'Global warming and terrorism', www.worldclimatereport.com/index.php/2005/07/12/global-warming-and-terrorism

3

Contributing to Resilience

Johan Rockström and Maria Schultz

Overwhelming evidence, including the Millennium Ecosystem Assessment (MEA), has clearly demonstrated that humans have changed ecosystems more rapidly and extensively over the past 50 years than in any other period in history (MEA, 2005). This has contributed to substantial net gains in human well-being and economic development but has largely and increasingly degraded most of the world's ecosystem services. The Economics of Ecosystems and Biodiversity (TEEB) estimates the annual welfare loss in the first ten years of the period 2000 to 2050 of ecosystem services from land-based ecosystems alone to be equivalent to approximately €50 billion under a 'business as usual' scenario (TEEB, 2008). This degradation of ecosystem services is increasingly jeopardizing human well-being, including possibilities of achieving the Millennium Development Goals (MDGs).

Active stewardship of biodiversity and ecosystem services is closely associated with the resilience of societies, in other words, their ability to remain in desired states and continue to develop in an era of increasing turbulence and global change. There is, undoubtedly, a high degree of drama when linking resilience with environmental change. We can no longer deny that negative environmental trends – such as loss of biodiversity, land and water degradation, air pollution and eutrophication – can erode resilience of social-ecological systems[1] and trigger regime shifts, causing abrupt and often irreversible change from desired to undesired states. We already see examples of such human-induced regime shifts or crossing of tipping points in lakes, coastal systems, degraded savannas and potentially in the Arctic. The insights of declining resilience in social-ecological systems together with the evidence of rapid global environmental change, shifts the agenda on sustainable development in a profound way, from a focus on minimizing environmental impacts to the active stewardship of social-ecological systems (Huitric et al, 2009).

Planetary Boundaries and Thresholds for Change

Anthropogenic pressures on the Earth-system have reached a scale where abrupt global environmental change can no longer be discounted. Recently, in a scientific effort to merge resilience theory with Earth-system science, a new planetary boundaries framework for global sustainability was introduced. This framework identified Earth-system processes and associated thresholds that, if crossed, could generate unacceptable environmental change from a global perspective. Nine planetary boundary processes were identified: climate change; rate of biodiversity loss (terrestrial and marine); interference with the nitrogen and phosphorous cycles; stratospheric ozone depletion; ocean acidification; global freshwater use; change in land use; chemical pollution; and atmospheric aerosol loading. Together these Earth-system boundaries provide a safe operating space for humanity and secure the future for human well-being and development (Rockström et al, 2009).

The rate of biodiversity loss is one of three boundary processes (together with climate change and the interference with the nitrogen cycle) where the analysis indicates that the safe boundary level has already been transgressed. Biodiversity loss interacts with several other planetary boundaries. For example, loss of biodiversity can increase the vulnerability of terrestrial and aquatic ecosystems to changes in climate and ocean acidity, thus reducing the safe boundary levels of these processes (Rockström et al, 2009).

From an Earth-system perspective, setting a safe boundary for biodiversity is difficult. Although it is now accepted that a rich mix of species underpins resilience, little is known quantitatively about how much and what kinds of biodiversity can be lost before this resilience is eroded. Nevertheless, there is a growing understanding of the importance of functional biodiversity in preventing ecosystems from tipping into undesired states when they are disturbed (Chapin et al, 2000; Purvis and Hector, 2000; Folke et al, 2004).

Hence, investing in resilience can be seen as insurance against future shocks. By safeguarding critical resources and ecological functions, the chances of 'riding through' shocks – such as extreme events – increase. This is of critical importance considering future uncertainty and limited understanding of the vulnerability generated by anthropogenic change.

There is a continuous need to understand the importance of biodiversity for decreased vulnerability in local to global systems. In many traditional cultures, resilience is a commonly practised knowledge. For instance, farmers safeguard food security and incomes by spreading their risks when planting many different kinds of crops and varieties along with home gardens. This diversity serves as a base and insurance for livelihoods (SwedBio, 2009).

The 2020 Strategic Plan for Biodiversity

At the 2002 World Summit on Sustainable Development (WSSD) in Johannesburg, South Africa, world leaders agreed to achieve, by 2010, a significant reduction of the current rate of biodiversity loss at the global, regional and national levels. This was seen as a contribution to poverty alleviation and to the benefits of all life on Earth. This target was also integrated into the Millennium Development Goals (MDGs). Collectively the world has not reached this target. In October 2010, at the tenth Conference of the Parties (COP-10) to the Convention on Biological Diversity (CBD) in Nagoya, Japan, the world's governments agreed upon a new strategic plan with measurable targets for 2020 and a vision for 2050.

This new strategy requires recognition of the dynamic interplay between biodiversity, ecosystem services and development in the context of rapid global environmental change. Needless to say this is a very challenging task. Success depends on an understanding of the ecological and social context in which targets can and should be set, as well as the governance context in which the targets should be implemented. Furthermore, it requires excellent communication outside the biodiversity community to make them understandable and relevant to human well-being and poverty reduction.

This chapter aims to stimulate discussion on the substantial challenges in meeting the biodiversity targets, including:

- Changing the current worldview;
- Improving the knowledge base;
- Good governance, flexible institutions, improved learning and dialogue between stakeholders and cultures; and
- Capacity-building, resource mobilization and innovative solutions.

Changing the Current Worldview

Despite the significant scientific consensus regarding the occurrence and impacts of global change, substantial action is still at a minimum. If the notion prevails that society is external to the environment rather than tightly interwoven with it, then actions to halt biodiversity loss and achieve sustainable development goals will have limited impact. Very few policy-makers are aware of the importance of biodiversity for resilient ecosystems and associated human communities. The pedagogical task of explaining the links between healthy ecosystems and the opportunities they provide to adapt to climate change is crucial (see Box 3.1 for examples). This is an urgent task for the world community as increasing scientific evidence warns of the potentially deleterious consequences if erosion of ecosystem services continues unabated.

Improving the Knowledge Base

More knowledge is needed to understand the importance of biodiversity within social-ecological systems. Furthermore, societies must become more resilient through the design of smarter social-ecological-based solutions and learning

Box 3.1 Potential of sustainable ecosystem management for climate change adaptation and mitigation

Sustainable ecosystem management can contribute to solutions to climate change, both for adaptation and mitigation (Trumper et al, 2009) and to disaster risk reduction.

More than a third of all greenhouse gas (GHG) emissions are related to agriculture and forestry.

The contribution from deforestation alone is approximately 20 per cent more than the entire transport sector (Stern, 2006).

Reducing deforestation could be a cost-effective way of reducing carbon dioxide emissions, if indigenous and local communities are involved, biodiversity is conserved and good governance and transparency is implemented, to secure resilience of social-ecological systems and long-term sustainability.

Examples include:

- Agriculture: Maintaining diversity of local varieties, crops and agricultural systems contributes to risk distribution, decreased vulnerability and increases the ability of the agricultural system to adapt. Increased levels of organic matter in soil contribute to increased harvests and improved ecosystem services, such as nutrient cycling and water retention, and also sequester substantial amounts of carbon dioxide in the soil.
- Forested mountain areas: These are important as water sources, but also for their capacity to absorb and moderate the consequences of flooding (and increased water flows from glacial melting).
- Coastal zones: Conservation of mangrove forests and coral reefs is a cost-efficient measure to protect coastal zones against weather-related catastrophes (such as storms, hurricanes and typhoons). It also benefits biodiversity and fisheries, since spawning grounds for fish are preserved, and it is favourable for tourism.
- Wetlands: These have a buffering effect (for example, against drought and flooding), as well as a rich species diversity. They also contribute to other ecosystem services such as removal of nitrogen from agricultural run-off (SwedBio, 2009).

from success stories. One example of the latter is agroforestry among smallholder farmers in the Rio Grande do Sul region of Brazil, which contributes to carbon sequestration, biodiversity conservation and higher income and better risk distribution for farmers (Lundberg and Moberg, 2009).

In terms of the knowledge base for the post-2010 biodiversity targets, there is a clear need for both new knowledge production and better assessment of the present knowledge. There is also a need to create arenas for understanding and learning from local and traditional ecological knowledge, as well as means of improving the compatibility of different forms of knowledge and promotion of interdisciplinary research between the natural and social sciences. Indicators have to consider ecosystem interactions and dynamics (internally and across scales). Several tools and processes already exist:

- Millennium Ecosystem Assessment (MEA) and Sub-Global Assessments (MEA, 2005; Ash et al, 2010): MEA was a report ordered by the United Nations (UN) and involved more than 1360 experts worldwide. MEA developed a framework for assessment of the consequences of ecosystem change for human well-being and linked assessments undertaken at local, watershed, national, regional and global scales. The primary goals of the MEA sub-global assessments were to meet the needs of decision-makers at the scale at which the assessments were conducted, to build capacity to undertake integrated assessments and to help develop and test methodologies for integrated multi-scale ecosystem assessments and approaches for integrating indigenous, local and traditional knowledge with scientific knowledge;
- Resilience assessments of social-ecological systems: based on two workbooks developed by the Resilience Alliance, targets both practitioners and scientists in the field of natural resource management (Resilience Alliance, 2007a, b);
- A framework for analysing sustainability of social-ecological systems developed by Elinor Ostrom (2009): focuses on both the natural resource in question and the institutional setting and provides an arena where researchers (and stakeholders) from different fields can share and join their understandings of the system;
- Ecosystem service valuation in the context of complexity, dynamics, thresholds and resilience (for example, TEEB, 2008): targets policy-makers, businesses and consumers in order to raise awareness of the value of biodiversity as well as the relative costs of inaction;
- Programme on Ecosystem Change and Society (PECS): the programme will advance research for governance and management of social-ecological systems to secure ecosystem services for human well-being;
- The Intergovernmental Science-Policy Platform on Biodiversity and Ecosystem Services (IPBES): the main focus of IPBES will be to perform assessments of knowledge on biodiversity and ecosystem services and identify key scientific information needed for policy-makers and catalyse efforts to generate new knowledge. IPBES will take an interdisciplinary and multidisciplinary approach that incorporates all relevant disciplines, including social and natural sciences. IPBES has the potential to provide information on the consequences and risks of continued loss of biodiversity, ecosystems and their services, to the top of the political agenda. As such, IPBES can help to bridge the gap between the closely interrelated agendas on climate, ecosystems and development.

Good Governance, Flexible Institutions, Improved Learning and Dialogue between Stakeholders and Cultures

There are many useful international agreements, for example, decisions under the CBD regarding sector integration, the ecosystem approach and decisions related to indigenous and local community rights and knowledge systems. What tends to be missing, however, is national implementation due to lack of financing, awareness and conflicting interests.

Many governance tools can be used to trigger ecosystem-smart and resilient development, such as green incentives as taxes and state budget lines for green development. The continued development in valuation of ecosystem services is an important task in order to demonstrate the importance of these services for human well-being, to decision-makers. When developing these tools and when implementing them social and equity aspects should be considered. The root causes of degradation of biodiversity and ecosystem services have to be addressed, for example, through elimination of perverse subsidies that hinder a sustainable ecosystem management. A better coherence between different policy areas – such as trade, development and environment – is needed.

There is also a need to work on a landscape level and further integrate biodiversity and ecosystem services aspects into sustainable agriculture, forestry and fishery.

An increasing world population will put further pressure on agriculture through rapidly rising food demand, which is expected to increase by 70 per cent by 2050, in order to eradicate hunger in a world population of 9 billion people by 2050 (FAO, 2006). The planetary boundaries analysis suggests that the degrees of freedom are limited for future expansion of crop land, freshwater use and extraction of phosphorus for food production; while extraction of nitrogen from the atmosphere (for fertilizers) and loss of biodiversity rates need to come down very rapidly. Food for a growing world population within the planetary boundaries will require a new 'planetary food revolution' that meets several challenges:

- On aggregate, contributes to remain within the safe operating space of the planetary boundaries, particularly with regards to the rate of loss of biodiversity, land-use change, freshwater use, interference with the global nitrogen and phosphorus cycles, and climate change.
- Increased efficiency and productivity of a broad spectra of ecosystem services, including – but not limited to – food, in the agricultural landscapes through the integration of innovative ways of managing land, water, crops and nutrients, while strengthening the resilience of agricultural landscapes by simultaneously increasing biodiversity.
- Optimizing efficiency in the food chain, from harvest through processing to consumption and recycling, food supply can increase with much less damage to the environment, similar to improvements in efficiency in the traditional energy sector (Nellemann et al, 2009).

- Increased attention to consumption patterns for reaching efficiency in resources use and to consider distribution, equity and rights aspects including the link between 'development' and the possibility of a growth in consumption by the world's poor majority.

There is a need to continue to improve ecosystem management tools such the integration of an ecosystem services approach in environmental impact assessments (EIAs), strategic environmental assessments (SEAs), to develop EIA/SEA regulation and compliance, and to continue to develop biodiversity/livelihood and ecosystem service indicators.

Tools for business include the *Corporate Ecosystem Services Review* (WRI, 2008), a methodology for corporate managers to proactively develop strategies for managing business risks and opportunities arising from their company's dependence and impact on ecosystems and human well-being. This is a step in the right direction to change the corporate sector's view of the environment and their interaction with it, and a reminder that this shift must occur across sectors.

A necessity in times of global change is awareness of uncertainty and surprise. This includes the will to experiment, innovate and learn within and between different knowledge systems and cultures. One of the root causes of our inability to make progress is that we live in a society where academia, media, law and politics cast complex problems as polar opposites (Costanza, 2010). To help build a shared vision of where our society wants to go and initiate a broad agreement about how to get there, we need an improved dialogue culture across cultures and interests.

Good contact between practitioners' reality, science and policy is required in order to enable successful policy decisions and recommendations and their further implementation at national and local levels. Public participation is necessary to set objectives for ecosystem service delivery in relation to stakeholder preferences and values (SwedBio, 2009).

In order to deal with uncertainty and surprise, we need to move away from steady-state approaches and blueprint solutions to more flexible governance structures and processes (Huitric et al, 2009). The implementation of IPBES will hopefully be an important mechanism for reducing uncertainty by identifying future trends, discovering emerging crises and thereby assist in preparing governments and regions in identifying response capacity to new issues.

The recent advancement in addressing such large-scale risks through the planetary boundary framework can contribute to operationalize the precautionary principle to allow for governance and management within a safe operating space.

Capacity-building, Resource Mobilization and Innovative Solutions

In the field of biodiversity and ecosystem services, funding at all levels is needed for institutional capacity-building, data collection, extension work and to create knowledge and learning interfaces. The European Union (EU) and developed countries play an important part in providing financial support. At the national level, it is crucial that the needs for biodiversity and ecosystem services are indicated in government budgets and Poverty Reduction Strategy Papers (PRSPs). The links between ecosystem services and economic and social development have to be made clear.

There are opportunities regarding innovative financial mechanisms, e.g. fiscal reforms. Any new and innovative funding mechanisms should be supplementary to financial mechanisms such as the Global Environment Facility (GEF) and other obligations for funding from developed countries. It is of utmost importance that development is democratically governed. For example, regulatory frameworks must be in place under which private sector investments should operate, to safeguard social and equitable development.

Conclusions

Social and ecological systems are interlinked. Solutions have to meet the needs of a growing population for ecosystem services such as food and water, while being climate and ecosystem smart and sustainable from a social as well as ecological perspective. Elinor Ostrom was awarded the 2009 Nobel Memorial Prize in Economic Sciences for 'her analysis of economic governance, especially the commons'. Common pool resources include forests, fisheries, oil fields, grazing lands and irrigation systems. Ostrom's work shows how humans interact with ecosystems to maintain long-term sustainable resource yields. She has shown how societies have developed diverse institutional arrangements for managing natural resources and avoiding ecosystem collapse. Her current work focuses on the multifaceted nature of human–ecosystem interactions and argues against any singular 'panacea' for individual social-ecological system problems. The choice for the Nobel Prize clearly demonstrates that there is an increased recognition that management of ecosystems is not just an environmental matter but also a development issue.

The above-mentioned planetary boundaries analysis reinforces earlier research on the urgent need for global action on bending the curve of biodiversity loss, not only as a way of preserving species on the planet, but as a strategic investment in human development for the future. The outcomes of 2010 – the year of biodiversity – and the agreements from COP-10 in Nagoya, such as the new strategic plan and its biodiversity targets, constitute important steps towards a sustainable development. We have a challenging road ahead in order to implement and revitalize important international commitments already taken, not just from the CBD but also others such as from the Framework Convention on Climate Change. One crucial opportunity in this respect is the upcoming United Nations Conference on Sustainable Development (UNCSD) in 2012 – also referred to as 'Rio+20', which is a follow-up to the UN Conference on Environment and Development (UNCED) 1992. Rio+20 seeks three objectives: securing renewed political commitment to sustainable development, assessing the progress and implementation gaps in meeting already agreed commitments and addressing new and emerging challenges. A timely and strategically important bridge between Nagoya and Rio is the 3rd Nobel Laureate Symposium in May 2011 in Stockholm. This will bring together some of the world's most renowned thinkers and experts on global sustainability and meet back to back with the UN's High-Level Panel on Global Sustainability, which will in its turn provide the framework for the Rio+20. We see these upcoming events as steps presenting a number of opportunities to turn the coming years into an era of real transformation. Let's take this opportunity and make real steps towards a more resilient society where the close linkages between social and ecological systems are understood and acted upon.

Note

1 Literature in the field includes Daniel Yankelovich's *The Magic of Dialogue: Transforming Conflict into Cooperation* (Simon & Schuster, 1999) and Deborah Tannen's *The Argument Culture: Moving from Debate to Dialogue* (Random House, 1998).

References

Ash, N., Blanco, H., Brown, C., Garcia, K., Henrichs, T., Lucas, N., Raudsepp-Hearne, C., Simpson, D. R., Scholes, R., Tomich T. P., Vira, B. and Zurek, M. (2010) *Ecosystems and Human Well-being: A Manual for Assessment Practitioners*, Island Press, Washington, DC

Chapin, F. S. III, Zavaleta, E. S., Eviner, V. T., Naylor, R. L., Vitousek, P. M., Reynolds, H. L., Hooper, D. U., Lavorel, S., Sala, O. E., Hobbie, S. E., Mack, M. C. and Díaz, S. (2000) 'Consequences of changing biodiversity', *Nature*, vol 405, no 6783, pp234–242

Costanza, R. (2010) 'The search for real, integrative solutions', *Solutions Journal*, vol 1, p1

FAO (Food and Agriculture Organization) (2006) *World Agriculture Towards 2030/2050*, Food and Agricultural Organization, Rome

Folke, C. S., Carpenter, S. R., Walker, H. B., Scheffer, M., Elmqvist, T., Gunderson, L. and Holling, C. S. (2004) 'Regime shifts, resilience, and biodiversity in ecosystem management', *Annual Review of Ecology, Evolution, and Systematics*, vol 35, pp557–581

Folke, C., Carpenter, S. R., Walker, B., Scheffer, M., Chapin, T. and Rockström J. (2010) 'Resilience thinking: Integrating resilience, adaptability and transformability', *Ecology and Society* vol 15, no 4, p20, available at www.ecologyandsociety.org/vol15/iss4/art20/

Huitric, M. (ed), Walker, B., Moberg, F., Österblom, H., Sandin, L., Grandin, U., Olsson, P. and Bodegård, J. (2009) 'Biodiversity, ecosystem services and resilience: Governance for a future with global changes', background report for the scientific workshop 'Biodiversity, ecosystem services and governance: Targets beyond 2010', Tjärnö, Sweden, 4–6 September 2009

Lundberg, J. and Moberg, F. (2009) 'Organic farming in Brazil: Participatory certification and local markets for sustainable agricultural development', Swedish Society for Nature Conservation, Albaeco, Stockholm

MEA (Millennium Ecosystem Assessment) (2005) *Ecosystems and Human Well-Being: Synthesis*, Island Press, Washington, DC

Nellemann, C., MacDevette, M., Manders, T., Eickhout, B., Svihus, B., Prins, A. G. and Kaltenborn, B. P. (eds) (2009) *The Environmental Food Crisis – The Environment's Role in Averting Future Food Crises: A UNEP Rapid Response Assessment*, United Nations Environment Programme, GRID-Arendal, Norway

Ostrom, E. (2009) 'A general framework for analyzing sustainability of social-ecological systems', *Science*, vol 325, no 5939, pp419–422

Purvis, A. and Hector, A. (2000) 'Getting the measure of biodiversity', *Nature*, vol 405, no 6783, pp212–219

Resilience Alliance (2007a) *Assessing and Managing Resilience in Social-Ecological Systems: A Practitioners Workbook*, www.resalliance.org/srv/file.php/208

Resilience Alliance (2007b) *Assessing Resilience in Social-Ecological Systems: A Scientists Workbook*, www.resalliance.org/srv/file.php/210

Rockström, J., Steffen, W.. Noone, K., Persson, Å., Chapin, F. S. III, Lambin, E. F., Lenton, T. M., Scheffer, M., Folke, C., Schellnhuber, H. J., Nykvist, B., de Wit, C. A., Hughes, T., van der Leeuw, S., Rodhe, H., Sörlin, S., Snyder, P. K., Costanza, R., Svedin, U., Falkenmark, M., Karlberg, L., Corell, R. W., Fabry, V. J., Hansen, J., Walker, B., Liverman, D., Richardson, K., Crutzen, P. and Foley, J. A. (2009) 'A safe operating space for humanity', *Nature*, vol 461, no 7263, pp472–475

Stern, N. (2006) *The Stern Review on the Economics of Climate Change*, Cambridge University Press, Cambridge

SwedBio (2009) *Contributing to Resilience: Results and Experience from the SwedBio Collaborative Programme 2003–2008*, Swedish Biodiversity Centre, Uppsala

TEEB (2008) *The Economics of Ecosystems and Biodiversity: An Interim Report*, European Commission, Brussels

ten Brink, P. (ed) (2011) *The Economics of Ecosystems and Biodiversity for National and International Policy Makers*, Earthscan, London

Trumper, K., Bertzky, M., Dickson, B., van der Heijden, G., Jenkins, M. and Manning, P. (2009) 'The natural fix? The role of ecosystems in climate change mitigation', United Nations Environment Programme World Conservation Monitoring Centre, Cambridge, www.unep.org/pdf/BioseqRRA_scr.pdf

UNEP (United Nations Environment Programme) (2009) *The Environmental Food Crisis: The Environment's Role in Averting Future Food Crises*, United Nations Environment Programme, GRID-Arendal, Norway

WRI (World Resources Institute) (2008) *The Corporate Ecosystem Services Review: Guidelines for Identifying Business Risks and Opportunities Arising from Ecosystem Change*, World Resources Institute, Washington, DC

4

Global Food Crisis: Biodiversity to Curb the World's Food Insecurity with a Focus on Indonesian Experience

Gusti Muhammad Hatta

Introduction

Food production worldwide has increased significantly during recent decades. However, continuous human population growth threatens food security. It is estimated that the world population will reach approximately 9 billion people by 2030, an increase of 3 billion. This significant increase demands 50 per cent more food (Glick, 2010). Yield increase of three major cereals (rice, maize and wheat) has been the major contributor in fulfilling the world food demand, while the extension of area planted with cereal crops plays a minor role and this mainly occurs in developing countries. Both agricultural intensification and expansion of crop area have limits and once these limits have been reached, food production would plateau.

Despite having the luxury of increasing food production, it is fully realized that competition for some natural resources – for example, water and land for agricultural systems – has become more prominent. Conversion of agriculturally productive land for housing and industrial purposes has been happening continuously. The effects of both climate change and this land conversion will intensify the pressure on harvested food crops, decreasing their long-term yields. In other words, environmental and industrial changes would have a significant impact on crops' productivity that would eventually threaten the state of our food security.

Biodiversity has a strong linkage with the stability and sustainability of an ecosystem because biodiversity offers genetic resources that could be used to increase human welfare, particularly to meet our demand for food. As millions of people worldwide are still malnourished, the utilization of biodiversity can be directed towards increases in production, distribution and accessibility of functional foods which assure good health. Indonesia is known to be among five countries with mega-biodiversity, and the following paragraphs will focus on the Indonesian experience in raising the national production of food crops and increasing food diversity to maintain sustainability and a more nutritious food supply. The expectation of collaborative research will be discussed as the main approach for the maintenance and utilization of biodiversity.

Agriculture and Biodiversity

Agriculture has gone through different eras, and one of the most significant changes in agricultural practices occurred during the agricultural revolution. The land use was operated close to its maximum capacity with additional high energy inputted from external sources. This effort had successfully increased the productivity of agriculturally important crops. Highly intensive use of land and high external input has raised some sustainability concerns related to soil erosion, depletion of non-renewable resources, damage to biodiversity and water availability. Indonesian agriculture has practised these intensive methods of food production.

Public concerns for reduced environmental quality due to agricultural activities have become greater, and demand for using resources in a more sustainable manner is getting stronger. Herman Daly (Meadows et al, 1992) proposed three criteria for how we must use the resources that will ensure sustainability. First, renewable resources must not be utilized at a greater pace than their regeneration rates; second, non-renewable resources must be used at a rate lower than the rate of their substitution made by renewable resources; and third, emission of pollutants must be lower than the ability of the environment to recycle them without any harmful effects.

Sustainable agriculture is common practice for farmers in many parts of Indonesia. For example, composting the remaining biomass after harvesting has been known to be useful for conserving and enhancing soil quality. Polyculture not only provides intermediate income for farmers but also increases and maintains diversity, decreasing the risk of disease thus reducing the need for chemical pesticides. Terracing and cover cropping in highland areas reduces soil erosion. These few examples give evidence that local wisdom existed before modern (high external energy inputs) agriculture was introduced. Modern agriculture offers advantages, particularly regarding the speed to which crops respond to the technology employed. It is a big challenge for public institutions to shift this trend and encourage the old wisdom to become the backbone of the Indonesian agricultural system in the future. Efforts toward this aim have been made but it needs to bring together all stakeholders invested in the agricultural sector.

Conserving the existing biodiversity is vital to ensuring a flourishing agricultural sector that can offer secure food production. Genetic resources are the basis for plant and animal breeding to create new varieties that are more adaptable to changing environments. Boehm and Morgan (1997) emphasize the importance of *in situ* and *ex situ* conservation of genetic resources. Taking the advantage of technology available in the country beyond its origin, a genetic resource could be well-managed and used for humanity. Linkages and collaborations between countries play an essential role in utilizing the maximimum benefit of the resources. Ensuring that collaboration is achieved in a fair relationship is vital to producing a just distribution to the resources' benefits.

Food Security in Indonesia

Food production

The Green Revolution helped Indonesia to attain self-sufficiency in rice during the mid-1980s. Yet, this achievement was not successfully maintained, mainly due to large population growth and a stagnation in rice productivity. Recently, the Indonesian government declared that self-sufficiency in rice was regained despite the fact that the population has reached 230 million. Food production has been growing at a high rate during the past five years, an average of 3.5 per cent per year during 2004–2007 and 4.8 per cent in 2008 (Food Security Council, 2009) with rice and maize acting as major contributors. This increase is higher than the rate of population growth, which is estimated at 1.3 per cent per year.

Domestic Indonesian cereal production is considerably sufficient for its people, particularly carbohydrate base cereals (rice and maize). However, the production of other grain, such as soybean, which is rich in protein, has been declining. As a result, Indonesia has been relying on the international market for meeting the domestic consumption of soybean. A further issue is the disparities between provinces and districts in cereal production. Districts in the Papua Province and some districts in several other provinces were deficient in cereal production (Food Security Council, 2009). However, this may not always mean that these regions experience a shortage of food, since some regions in Indonesia do not use rice as a staple food in their daily diet. If this deficiency is a case of local production being lower than local demand, distribution within the country will solve the problem. The deficient areas have mostly experienced problems related to climatic conditions, unsuitable soil or natural disasters, such as flooding and drought.

Java is one of several cereal-producing islands. However, the area of agricultural land has continuously decreased due to the conversion of land for housing and factories. To compensate for the loss in production, new areas have had to be opened up outside Java. Forest areas in Sumatera, Kalimantan and Papua are in excess of the recommended forested area based on agro-ecological zoning (Food Security Council, 2005). This implies that current forested areas could be converted for agricultural purposes. These conversions must be well-planned in order to successfully promote food crop cultivation from unexploited lands in those islands. Less labour-intensive agricultural or agro-forestry systems may have a better chance of success than labour-intensive systems such as seasonal crops.

Food access

Although the availability of food at the national level is adequate, in 2008, 15.42 per cent of the population (34.96 million people) still lived under the national poverty line (Purchasing Power Parity, US$1.55 per day). They mostly (about 64 per cent) lived in the rural areas, especially on Java Island (57 per cent). The number of poor people increased slightly from 34.01 million people in 1996 (Food Security Council, 2009). There are a number of reasons accounting for limitation to food access, including unemployment, lack of permanent job, poverty or low income. Reducing the number of poor remains a challenging task, and understanding the root of poverty will be the basis for developing effective programmes.

A 2 per cent decrease of open unemployment rate was recorded between 2003 and 2007, but this did not significantly reduce the number of poor people. Poverty is most likely linked to lack of infrastructure. Some Indonesian villages (12 per cent) had no access to the main roads linked by four-wheeled vehicles, and almost 10 per cent of the households were not supported by the availability of electricity. Most of these conditions occurred in eastern Indonesia. Furthermore, the geographical conditions of that region are mostly not suitable for high yield crops. Improvement to the infrastructure of the poor regions is essential because it will reduce the cost of transporting surplus agricultural produce from rural to urban areas. This change is then expected to create better purchasing power for different foods. Provision of productive and durable employment in the poor regions is another main strategy in reducing poverty because it will limit the temporary or permanent migration of people to regions with better employment conditions.

Food utilization

The total energy and protein intake increased 3.3 per cent during 2002–2007, and the food basket became more variable. Consumption of cereals and tubers declined and was substituted by greater intake of animal products, vegetables, fruits and oil that contained more protein and vitamins. The daily energy intake in 2007 averaged 2050 Kcal/person/day, which exceeded the national recommended daily allowance (RDA) of 2000 Kcal/person/day. Similarly, the protein consumption reached 56.25g/person/day, and it was higher than its RDA of 52.0g. Improvements in consumption patterns occurred at all levels of income, even though the lowest monthly per capita expenditure (MPCE) experienced the lowest rate of increase. People with the lowest MCPE were only able to consume 69 per cent and 67 per cent of the national RDA for energy and protein respectively (Food Security Council, 2009). Clearly, more can be done to help the poorest who have not seen as great a benefit in their relative energy and protein intake.

Applying World Health Organization (WHO) criteria, 18.4 per cent and 36.8 per cent of children aged five and below in 2007 fell into the category of underweight and stunting respectively. Although the number of underweight was considerably high, it showed that the target of the Medium Term Development Plan for Nutrition Programme (20 per cent) and MDGs 2015 (18.5 per cent) have been surpassed. More than that, disparities remain among provinces and districts. Nineteen provinces had a higher number of underweight children than that of the national average. This figure correlates with the data on food production and food access. In other words, most of the underweight children live in the regions with low production and food access.

To address the problems of food consumption and nutritional status, the Indonesian government has implemented three major policies (Food Security Council, 2009):

1. Encouraging society to diversify the source of foods emphasizing the usage of local foods to support a healthy and productive life;
2. Increasing the awareness and practices of people to consume a greater variety of food that is balanced in nutritional values and safe;
3. Supporting the development of food processing technology, based on local non-rice food to add value and raise its social status. These policies have been translated into a road map starting from 2007 to 2015. Social engineering through education and awareness campaigns is one of the main strategies, among others listed, to produce a localized and diversified food consumption pattern that adequately meets the nutritious and safety needs of its people.

Evolution in agriculture toward sustainability

Increasing food production is occasionally translated into practice as a way of investing more external inputs for agriculture. Considering the limitations of renewable and non-renewable resources, agricultural practices that consume less energy are becoming popular. This means that the success of this transformation will be greatly dependent upon our understanding of biodiversity and how the ecosystem works, and how we capitalize on it to address the issues related to food security and food safety. Developing access to high-yielding seeds that are adaptable to the local environment is a further factor that will determine the extent of the food crisis.

Agricultural practices in Indonesia have changed somewhat and they can be categorized into three major eras. The first era was before the independence of Indonesia (prior to 1945). This era was characterized mainly by low energy input and by the practising of local wisdom and indigenous knowledge for farm management. The second era started in the late 1960s. The problem of food availability was evident, and this triggered the government to launch different programmes of rice intensification, starting from *Bimbingan Massal* (BIMAS) to *Supra Intensifikasi Khusus* (Supra Insus). These programmes required intensive farm management and investment of external energy, such as chemical fertilizers and pesticides. Following increasing awareness of the side effects of chemical inputs on the environment and human health, a new paradigm was adopted. This era was signified by the issuance of a Presidential Decree in 1986, which mandated the implementation of integrated pest management (IPM) and the withdrawal of 57 insecticides for use in rice. In collaboration with the Food and Agriculture Organization (FAO), the government initiated a national programme of IPM in 1989. Farmers received training through field schools, emphasizing the role of farmers in making decisions for their farm based on ecosystem analysis. Millions of farmers have participated in such programmes, and alumni of this programme became trendsetters during the agricultural transition towards a more eco-friendly system.

In the development of agriculture, there have been several variants of practices that have had more reliance on the use of natural resources in a sustainable manner. These include low-input sustainable agriculture, integrated farming systems, organic farming, biological intensive pest management and system rice intensification. Science and technology could complement what nature has provided, and the development of technology should enrich and be in line with this transitional transformation. For example, biotechnology could function in line with the maintenance and utilization of genetic resources.

The era of globalization has increased the movement of agricultural products. The export and import of the products is regulated by international agreement. This brings the opportunity for every country to utilize the international market. At the same time, international trading also brings some risk associated with the introduction of invasive alien species (IAS), of which Indonesia has some experience. IAS have caused not only economic losses but also environmental problems. They can create ecological displacements due to competition with the domestic species that eventually endangers the biodiversity of the existing ecosystem. Therefore, the maintenance of biodiversity is not only a matter of avoiding ecological disturbance due to domestic agricultural activities but also preventing the entrance of IAS.

Future Direction

Research

Conservation of genetic resources is the foundation of plant and animal breeding, which increases the opportunity to avoid or minimize the risk of food crises in the future. Investment for research related to conservation should include methods of conservation, determination of the value of biodiversity and the impact of adopted agricultural technologies on biodiversity (Boehm and Morgan, 1997; Fraleigh, 1997; Reeves et al, 1997). Discussions have, for some time, highlighted the existence of health benefits associated with crops, beyond just simply the nutrition they provide (Hasler, 2002).

Having no doubt regarding the importance and urgency of conserving genetic resources, countries with different diversity indices may take different research pathways. Indonesia, as one of the five most populous and most diverse countries in the world, would prioritize the improvement of food production to fulfil the domestic requirement, while maintaining and increasing the quality of natural and environmental resources to ensure sustainability.

New model of partnership

The loss of biodiversity has significant consequences from economical and ecological perspectives. To avoid increased levels of starvation, the increase in food production needs to match the growth of world population. Global food crisis will most probably be overcome through collaboration between developed and developing countries. Such collaboration should ensure fairness and equitable sharing of benefits. Educational programmes must be an integral part of the partnership being created. The ability for developing countries to understand and contribute to the critical thinking on the functions of biodiversity would help to curb the world's food insecurity.

Conclusions

Agriculture keeps fulfilling its role in providing foods, and its capacity is considerably determined by our success in plant and animal breeding using existing genetic resources. Therefore, the conservation of biodiversity is a must since our future relies on it. Indonesia has undergone many different experiences in relation to food security. At the national level, food production is adequate and well-planned distribution will ensure the movement of the product from surplus to deficient regions. There remains a challenge regarding nutritional values because a significant proportion of the population consumes a mostly carbohydrate-based diet without having the buying power to fulfil nutritional balance. Research needs to be conducted collaboratively to speed up the transfer of skills and technology, and match the demand for foods worldwide. Future agriculture should balance and synergize the need for food and the need for medicinal products to increase health.

Acknowledgements

The author wishes to thank the Indonesian Food Security Agency and Y. Andi Trisyono for their valuable inputs and data for the early draft.

References

Boehm, M. and Morgan, B. (1998) 'Biodiversity conservation for sustainable agroecosystems workshop', in R. W. F. Hardy, J. B. Segelken and M. Voionmaa (eds) *Resource Management in Challenged Environments*, NABC Report 9, National Agricultural Biotechnology Council, Ithaca, NY, http://nabc.cals.cornell.edu/pubs/nabc_09.pdf

Food Security Council (2005) *A Food Insecurity Atlas of Indonesia*, Department of Agriculture of the Republic of Indonesia and World Food Programme of the United Nations, Jakarta

Food Security Council (2009) *A Food Security and Vulnerability Atlas of Indonesia*, Department of Agriculture of the Republic of Indonesia and World Food Programme of the United Nations, Jakarta

Fraleigh, B. (1997) 'Issues in agricultural biotechnology and biodiversity for sustainable agroecosystems', in R. W. F. Hardy, J. B. Segelken and M. Voionmaa (eds) *Resource Management in Challenged Environments*, NABC Report 9, National Agricultural Biotechnology Council, Ithaca, NY, pp105–112, http://nabc.cals.cornell.edu/pubs/nabc_09.pdf

Glick, H. (2010) 'Food security: Innovation in production technology', workshop on Indonesia–US Partnership: Agricultural Innovation and Investment, 2 March 2010, Jakarta

Hasler, C. (2002) 'Where do functional foods fit in the diet?', in A. Eaglesham, C. Carlson and R. W. F. Hardy (eds) *Integrating Agriculture, Medicine and Food for Future Health*, NABC Report 14, National Agricultural Biotechnology Council, Ithaca, NY, pp137–141, http://nabc.cals.cornell.edu/pubs/nabc_14.pdf

Meadows, D. H., Meadows, D. L. and Randers. J. (1992) *Beyond the Limits: Confronting Global Collapse, Envisioning a Sustainable Future*, Chelsea Publishing Co, Vermont

Pangan, B. K. (2009) *Pedoman Umum: Percepatan Penganekaragaman Konsumsi Pangan dan Gizi (P2KPG)* (General Guideline: Increasing the Diversity of Food Consumption and Nutrition), Department of Agriculture of the Republic of Indonesia, Jakarta

Reeves, T., Pinstrup-Andersen, P. and Pandya-Lorch, R. (1997) 'Food security and the role of agricultural research', in R. W. F Hardy, J. B. Segelken and M. Voionmaa (eds) *Resource Management in Challenged Environments*, NABC Report 9, National Agricultural Biotechnology Council, Ithaca, NY, pp97–102, http://nabc.cals.cornell.edu/pubs/nabc_09.pdf

5

Coral Reefs

Wolfgang Kiessling and Georg A. Heiss

Coral Reefs Today

Biodiversity, distribution, and ecosystem services

Tropical coral reefs are extremely complex ecosystems and true diversity hot spots in the oceans. Only a small fraction of the diversity is made up by stony corals and coralline algae, which build the reefs. The bulk is provided by fish and invertebrate animals dwelling in and around the reefs. The combined species richness of coral reefs has been tabulated to be close to 100,000 (Reaka-Kudla, 1997), but true diversity could well be ten times as great. That 4–5 per cent of all described living species are associated with coral reefs is astounding: reefs occupy only a small fraction of the oceanic habitats (less than 0.1 per cent) let alone the entire Earth. Also, reefs lack insects that make up the bulk of biodiversity in tropical rainforests, the terrestrial equivalents of coral reefs. Reef diversity varies substantially with geographic region. Highest diversity is found in the 'coral triangle' in the tropical west Pacific, lowest diversity is observed in the Caribbean.

Tropical coral reefs only occur between 34°N and 32°S in well-lit shallow water. Their ecological requirements can be well-established by looking at a global distribution map. Reefs are heavily concentrated in ocean interiors and at the western margins of oceans. This is due to cool and nutrient-rich currents moving along the western margins of north–south oriented continents. The ability of tropical reef corals to thrive in oligotrophic settings is clearly linked to the coral–algal symbiosis. Corals take up most of their food from the photosynthetic products of algae that they host in their tissues, whereas the algae receive carbon dioxide and inorganic nutrients from the coral animal. Symbiosis also enhances growth rates of corals by the removal of carbon dioxide for algal photosynthesis.

A completely different type of coral reefs gained attention only in the past two decades or so. These are cold-water reefs known from the Polar Circle to the Southern Ocean. The corals building these reefs do not harbour symbionts and depend entirely on external feeding on zooplankton and particulate organic matter. As they do not require light, they thrive in deeper waters of 200–1000m. The diversity of corals in these reefs is much lower than even in moderately diverse tropical settings. Nevertheless, cold-water coral ecosystems are biodiversity hot spots in the deep sea with hundreds of associated species, including fish and invertebrate animals (Roberts et al, 2006).

Coral reefs provide many important ecosystem services. Thanks to very efficient nutrient recycling in reefs, a large biomass of fish and other animals can be maintained in regions that are otherwise oceanic deserts. Fisheries in coral reefs provide a major food source. The aesthetic and recreational value of coral reefs attracts tourists. Reefs protect coastlines from wave action and erosion and provide building materials. Due to the intense competition, reef organisms developed a large suite of chemical compounds for defence and attack. Many of these substances are expected to be useful for new drugs. Snails and sponges have already delivered medicines for the treatment of cancer and melanoma but the potential of reefs for pharmacy is still underexplored. The genetic treasure of reefs is perhaps the most direct service that is offered by biodiversity. The direct economic service of coral reefs has been estimated to be US$30 billion per year (Cesar et al, 2003) but this number is very conservative as it excludes cold-water reefs and uncovered medical treasures.

Current conditions and the shifting baselines syndrome

Tropical reefs are clearly in decline and they have been for a long time. The *Status of Coral Reefs of the World: 2008* report (Wilkinson, 2008) estimates that 19 per cent of the global reef area is destroyed or non-functional, another 15 per cent is under imminent threat and will be lost within the next 10–20 years. It also states that 20 per cent of the reefs will be lost in 20–40 years, if the current pressure remains. Less than half of all reefs are under 'low threat' by human activities. Reef degradation caused by man can be traced back to times when humans were still hunter-gatherers, and has accelerated in modern times (Pandolfi et al, 2003). Large animals (fish, turtles and mammals) were affected first, whereas reef corals were little affected until recently. Among the regions with sufficient historical data, the best-protected reefs in the world on the Great Barrier Reef are still in the best condition. Some reefs surrounding Pacific islands and atolls appear to be in a semi-pristine state (Spalding et al, 2001), but the lack of historical data prevents a conclusive statement. Caribbean reefs are more than 50 per cent along their way from being in a pristine state to being ecologically extinct.

Consistent criteria and historical data are important to circumvent the shifting baselines syndrome, which manifests that, due to the lack of objective standards, every generation tends to consider a 'pristine' baseline as what they observed in their youth. Another side of the shifting baselines syndrome is that some threats might be exaggerated when they have never been encountered. Examples are widespread coral diseases, coral bleaching, the outbreaks of coral-eating starfish and the decline of herbivorous sea urchins. In the absence of solid historical data, these threats cannot be put in context and actions could be costly and in vain.

Lessons from the Past

Deep time

Reefs, broadly defined as laterally confined limestone structures built by the growth or metabolic activity of sessile benthic aquatic organisms (Kiessling, 2003), have been pertinent features of the oceans almost since the origin of life (Allwood et al, 2006). Corals and sponges, together with calcareous algae and calcifying bacteria have been the most prolific reef-builders for 550 million years. Many important lessons can be learned from the fossil record: first, the history of reefs is very volatile. Long periods of reduced reef growth are interrupted by relatively protracted episodes of enormous reef production (Kiessling, 2009). This observation has two implications: (a) even without anthropogenic influence, reefs experienced a great deal of natural variation; and (b) there must be tipping points for both the waxing and the waning of reefs. Thresholds, known as phase shifts (Nyström et al, 2008), have also been observed on much shorter timescales.

Second, five significant reef crises are evident in the past 500 million years. Three of these reef crises match with global mass extinctions, but two were associated with only moderate extinctions in the oceans (Kiessling et al, 2010). Four of the five reef crises were probably associated with rapid global warming and ocean acidification; the fifth can be linked to enhanced nutrient delivery from land. This suggests that factors that are held responsible for the current reef decline were also pertinent in the ancient past.

Third, ecological changes in reefs tended to be reduced when reef diversity was high and vice versa (Kiessling, 2005). This implies that diversity–stability relationships can act over very long intervals of time and affirms the great importance of maintaining biodiversity.

Finally, reefs have been shown to generate biodiversity and many species to other marine habitats (Kiessling et al, 2010). Coral reefs are thus not just ecological attractors of many species, but evolutionary rates are greater than anywhere else in the oceans. Destroying reefs thus means that recovery from the current biodiversity crisis will be delayed in the entire ocean.

Shallow time

The modest ecological response of coral reefs to the large climatic variations during the Pleistocene epoch (2.6 million years to 12,000 years before the present) has been attested in several studies (Pandolfi, 1999; Tager et al, 2010). Glacial–interglacial cycles driven by changes in the Earth's rotation and orbit resulted in substantial climate change and fluctuations in ice volumes in high latitudes, which caused massive changes in global sea–level. The stability of reefs in terms of ecology and negligible extinction during the Pleistocene are surprising. The limited response to cooling could be explained by the slow pace of cooling. But why was there so little effect of the sharp warming intervals? The answer may lie in the ameliorated climate change in the tropics and in the fact that critical tipping points were not surpassed in the Pleistocene warming intervals. In summary the geological record tells us that global change may only become a severe issue under extreme boundary conditions.

Current Threats

Threats to coral reefs are usually separated into global and regional. Their combined effect is that one third of all reef-building corals are currently considered to be at elevated extinction risk (Carpenter et al, 2008).

Global threats

The most eminent global threats are global warming and ocean acidification (Hoegh-Guldberg et al, 2007; Veron et al, 2009), which are both caused by the anthropogenic increase of greenhouse gases (GHGs), in particular carbon dioxide. In the global warming scenario, it is less the increase of average sea-surface temperatures but the increased frequency and severity of extreme temperature events that is thought to be harmful to reefs. Warming pulses induce coral bleaching, the phenomenon of stress-induced expulsion of symbiotic algae, which may lead to the death of corals. Adaptation and acclimation may increase the resistance of recovering reefs to future bleaching (Baker et al, 2004), but it seems unlikely that corals and their symbionts are able to adapt or acclimate rapidly to short, sporadic warming (Hoegh-Guldberg, 1999). Even if some species successfully adapt, levels of biodiversity will certainly be reduced.

Some reef corals may survive in higher latitudes (Precht and Aronson, 2004; Greenstein and Pandolfi, 2008). However, while migration may prevent global extinction, it does not prevent local extinction and reef destruction in areas affected by warming. A large fraction of anthropogenic carbon dioxide emissions is absorbed by the surface ocean, causing a lowering of pH and thereby a reduction of the super-saturation of seawater with respect to calcium carbonate. This ocean acidification seems to be especially harmful for reef corals and eventually entire reef systems. Growth rates of reef corals are already reduced relative to pre-industrial rates (De'ath et al, 2009). There is no indication that corals can adapt to ocean acidification, as calcification consistently decreases with decreasing pH (Kleypas and Langdon, 2006).

Regional threats

A large suite of regional factors have been identified to cause severe damage for coral reefs. Nearly all of them are attributable to human population growth and increased migration to coastal areas. These are in order of importance: overfishing, pollution and physical destruction.

Overfishing has severe consequences for the entire reef ecosystem, because it selectively affects large species, which often fulfil important ecological services. Examples are apex predators such as sharks and groupers and herbivorous fish such as parrotfish. Herbivorous fish are often key to prevent algal overgrowth of corals (Knowlton and Jackson, 2008) and predators are known to be critical for maintaining food web complexity and diversity (Paine, 1966). Overfishing is often associated with destructive fishing methods with explosives or poison. In addition to killing much more than is usable, explosives also destroy the physical structure of the reef, resulting in the loss of habitat and erosion of the reef. Poison is usually applied in the extremely lucrative live reef fish food trade, which is often controlled by organized crime.

The importance of pollution cannot be overstated. Corals are not only able to thrive in nutrient-depleted waters, they require them. Nutrient input is associated with a proliferation of nutrient-opportunistic algae, both in the water column and on the sea floor. The effects are that sunlight cannot penetrate deeply into the water column reducing the efficiency of the symbiosis and that fleshy algae overgrow corals to eventually kill them. Sediment input reduces light availability and requires energy from the coral polyps to clean the tissue; energy that is lost for other essential life functions. Thus any sort of pollution, be it enhanced nutrient input from agriculture, industrial or touristic waste or terrigenous sediment, is very harmful to reefs.

In the absence of other building materials, coral reefs are often excavated for construction work or used as building grounds for airports, highways or hotels. Physical damage is also caused by ship groundings, military activities, snorkellers and divers. Physical destruction by bottom trawling is severe in cold-water reefs where recovery is measured in hundreds to thousands of years (Roberts et al, 2006).

Consequences

Increasing insecurity

Stressed reefs may respond quite differently to natural disturbances than healthy reefs. A disturbance such as a big storm that previously triggered the renewal and development of reefs might, under stress, become an obstacle to development (Nyström et al, 2000).

The erosion of biodiversity of reefs is diminishing livelihood opportunities in many tropical countries, resulting in conflict and law and order problems (Govan, 2009). This will exacerbate problems associated with poverty and result in increased pressure on reef ecosystems and other natural resources.

Consequences for humanity

- Overfishing and reef destruction are leading to a reduction of the productivity of fisheries and causing economic and social problems in many reef countries. The effects will be greatest in the coral triangle where more than 150 million people rely on seafood.

- International trade is driving destructive fishing practices and unsustainable harvests from coral reef ecosystems, reducing the value of coral reefs to local communities and prospects for long-term sustainable use (Moore and Best, 2001)
- Loss of coastal protection is leading to coastal erosion, more flooding from storm surges and rainstorms, with impacts on coastal infrastructure, physical reduction of available land and the potential reduction in freshwater availability.
- The combined effects of sea-level rise and reef destruction poses severe threats to low-lying islands with the risk of complete loss of habitable land and freshwater availability.
- Even in areas where ecosystem functioning is not affected and the reefs remain intact, a loss of biodiversity will reduce the aesthetic value of coral reefs and probably cause a severe reduction of tourism in countries that strongly depend on this income.
- Declining biodiversity on reefs will reduce the potential to discover new medically active compounds.

Future scenarios

Hoegh-Guldberg et al (2007) have identified a 2°C global warming and a 480ppm concentration of carbon dioxide in the atmosphere as a critical threshold for the stability of the world's coral reefs. There are model calculations that suggest that a doubling of carbon dioxide concentration relative to pre-industrial levels, that is, to 560ppm, will mean the end of reef growth in all oceans (Silverman et al, 2009). As the saturation state of seawater is lower in deeper waters, cold-water reefs are expected to be even more strongly affected by ocean acidification (Guinotte et al, 2006).

Response Options

The maintenance of the status quo is clearly insufficient to reverse the trajectories of reef decline (Pandolfi et al, 2003). But we have to be realistic regarding response options. Ideally we would want to target all potential threats simultaneously but, as money is an issue, we have to rank response options by their acuteness. From what has been stated above, we deduce that acting regionally is probably of highest priority, although the threshold effects by global change have to be considered.

Global actions

The greatest threat to long-term sustainability of reefs is probably carbon dioxide. Reduction of GHG emissions should continue to be a major effort, especially in developed countries and fast-growing economies. Large-scale carbon dioxide removal options such as ocean fertilization still require proof of their effectiveness and have high risks of serious side effects of unknown extent. Bottom trawling for deep-water fish should be banned in international waters as it causes severe damage to cold-water reefs.

Regional and local actions

Most countries with tropical reefs have limited financial capacity to implement sufficient management interventions. Developed countries must increase their efforts to build partnerships and commit to substantially increase financial contributions for local management, which can support the resilience of reefs to global change.

Overfishing and pollution should be the major concern. Marine protected areas (MPAs) are a good option, especially in regaining fish diversity and richness (Halpern, 2003), but are also generally effective in reducing or preventing coral loss (Selig and Bruno, 2010). The harvest of herbivorous fish must be reduced to sustainable levels (Hughes, 2007) and top predators must be protected. Although it may take decades for fish populations to recover to pristine states (McClanahan et al, 2007), the fact that they do recover when no-take areas are defined and enforced does suggest that MPAs are good investments. On a local basis, community-based resource management approaches, set up by communities and non-governmental organizations (NGOs), have been successful in the South Pacific (Govan, 2009). Because they involve the people who are directly affected in the management process, there is a better chance for the long-term acceptance of sustainable management schemes. Well-connected transnational networks of large MPAs are essential to maintain connectivity and high levels of biodiversity, and protected areas need to be permanent to enable full recovery. Land-based pollution from agriculture, human settlements and industry must be minimized and improved cooperation of authorities dealing with coastal and land-based issues is necessary.

Suggestions to Governments

- **Create and enforce marine parks:** There is an urgent need for more, larger, better managed and better connected MPAs as the backbone of reef protection. The focus on protected areas should not lead to neglect of the remaining, much larger reef areas that are not protected. Here, efficient management schemes with the involvement of the local population are necessary, including the establishment and control of fishing quota and the creation of options for alternative income.
- **Support sustainable tourism:** The increasing importance of tourism should be recognized as a chance to increase the value of healthy coral reefs for national economies as well as for the local population. Sustainable tourism, avoiding the negative impacts of mass tourism, can bring high revenue and create incentives for the protection and responsible use of reef resources. A portion of the income created by tourism should flow back into reef protection, for example, for building more efficient sewage plants.
- **Facilitate reef research:** Currently our understanding regarding the response to combinations of different stressors acting simultaneously and often synergistically is only in its infancy. Complex interactions between climate, physical and chemical conditions, marine ecosystems and fishery must be

better understood to allow reliable predictions of how marine systems respond to global change and management interventions. Increased and stable funding of research as well as access for scientists to coral reef ecosystems is necessary to allow research in support of effective management.

- **Seek partners in pharmaceutical exploitation:** Reef countries should be encouraged to build partnerships with agencies and companies active in pharmaceutical research with fair conditions. Concerns about the inadequate use of genetic resources from reefs must be addressed in international agreements. Governments should support current negotiations on building an international regime to promote and safeguard the fair and equitable sharing of benefits arising from genetic resources within the framework of the Convention on Biological Diversity (CBD) Access and Benefit Sharing.
- **Improve ecological and socio-economic monitoring:** There is a severe lack of public funding for long-term monitoring programmes, both on national and international levels. Governments and organizations should increase their funding in this area. This involves all levels of observation, from community-based volunteer programmes such as Reef Check (Carpenter et al, 2008), monitoring programmes by park authorities or national authorities, detailed scientific studies about the effectiveness of tools, governance and design of MPAs to the synopsis in publications such as *Status of Coral Reefs of the World* (Wilkinson, 2008).
- **Improve legislation:** Recent accidents involving cargo ships and oil-drilling platforms have again highlighted the lack of adequate legislation and enforcement in marine environments. Control of land-based and ship-based pollution, regulation of coastal construction, mandatory environmental impact assessments (EIAs), legislation and enforcement of safety of shipping and offshore drilling activities must be improved on national and international levels.

References

Allwood, A. C., Walter, M. R., Kamber, B. S., Marshall, C. P. and Burch, I. W. (2006) 'Stromatolite reef from the Early Archaean era of Australia', *Nature*, vol 441, no 7094, pp714–718

Baker, A. C., Starger, C. J., McClanahan, T. R. and Glynn, P. W. (2004) 'Coral reefs: Corals' adaptive response to climate change', *Nature*, vol 430, no 7001, p741

Carpenter, K. E., Abrar, M., Aeby, G., Aronson, R. B., Banks, S., Bruckner, A., Chiriboga, A., Cortés, J., Delbeek, J. C., DeVantier, L., Edgar, G. J., Edwards, A. J., Fenner, D., Guzmán, H. M., Hoeksema, B. W., Hodgson, G., Johan, O., Licuanan, W. Y., Livingstone, S. R., Lovell, E. R., Moore, J. A., Obura, D. O., Ochavillo, D., Polidoro, B. A., Precht, W. F., Quibilan, M. C., Reboton, C., Richards, Z. T., Rogers, A. D., Sanciangco, J., Sheppard, A., Sheppard, C., Smith, J., Stuart, S., Turak, E., Veron, J. E. N., Wallace, C., Weil, E. and Wood, E. (2008) 'One-third of reef-building corals face elevated extinction risk from climate change and local impacts', *Science*, vol 321, no 5888, pp560–563

Cesar, H., Burke, L. and Pet-Soede, L. (2003) *The Economics of Worldwide Coral Reef Degradation*, Cesar Environmental Economics Consulting and WWF-

Netherlands, Arnhem and Zeist, The Netherlands, http://pdf.wri.org/cesardegradationreport100203.pdf

De'ath, G., Lough, J. M. and Fabricius, K. E. (2009) 'Declining coral calcification on the Great Barrier Reef', *Science*, vol 323, no 1, pp116–119

Govan, H. (2009) *Status and Potential of Locally-managed Marine Areas in the South Pacific: Meeting Nature Conservation and Sustainable Livelihood Targets Through Wide-spread Implementation of LMMAs*, Coral Reef Initiatives for the Pacific, New Caledonia, and Secretariat of the Pacific Regional Environmental Programme, Samoa, www.sprep.org/att/publications/000646_LMMA_Report.pdf

Greenstein, B. J. and Pandolfi, J. M. (2008) 'Escaping the heat: Range shifts of reef coral taxa in coastal Western Australia', *Global Change Biology*, vol 14, no 3, pp513–528

Guinotte, J. M., Orr, J., Cairns, S., Freiwald, A., Morgan, L. and George, R. (2006) 'Will human-induced changes in seawater chemistry alter the distribution of deep-sea scleractinian corals?', *Frontiers in Ecology and the Environment*, vol 4, no 3, pp141–146

Halpern, B. S. (2003) 'The impact of marine reserves: Do reserves work and does reserve size matter?', *Ecological Applications*, vol 13, no sp1, pp117–137

Hoegh-Guldberg, O. (1999) 'Climate change, coral bleaching and the future of the world's coral reefs', *Marine & Freshwater Research,* vol 50, no 8, pp839–866

Hoegh-Guldberg, O., Mumby, P. J., Hooten, A. J., Steneck, R. S., Greenfield, P., Gomez, E., Harvell, C. D., Sale, P. F., Edwards, A. J., Caldeira, K., Knowlton, N., Eakin, C. M., Iglesias-Prieto, R., Muthiga, N., Bradbury, R. H., Dubi, A. and Hatziolos, M. E. (2007) 'Coral reefs under rapid climate change and ocean acidification', *Science*, vol 318, no 5857, pp1737–1742

Hughes, N. C. (2007) 'The evolution of trilobite body patterning', *Annual Review of Earth and Planetary Sciences*, vol 35, pp401–434

Kiessling, W. (2003) 'Reefs', in G. V. Middleton (ed) *Encyclopedia of Sediments and Sedimentary Rocks*, Kluwer Academic Publishers, Dordrecht, The Netherlands, pp557–560

Kiessling, W. (2005) 'Long-term relationships between ecological stability and biodiversity in Phanerozoic reefs', *Nature*, vol 433, no 7024, pp410–413

Kiessling, W. (2009) 'Geologic and biologic controls on the evolution of reefs', *Annual Review of Ecology, Evolution, and Systematics*, vol 40, pp173–192

Kiessling, W., Simpson, C. and Foote, M. (2010) 'Reefs as cradles of evolution and sources of biodiversity in the Phanerozoic', *Science*, vol 327, no 5962, pp196–198

Kleypas, J. A. and Langdon, C. (2006) 'Coral reefs and changing seawater carbonate chemistry', in J. T. Phinney, O. Hoegh-Guldberg, J. Kleypas, W. Skirving and A. Strong (eds) *Coral Reefs and Climate Change: Science and Management*, AGU Monograph Series, Coastal and Estuarine Studies, vol 61, American Geophysical Union, Washington, DC, pp73–110

Knowlton, N. and Jackson, J. B. C. (2008) 'Shifting baselines, local impacts, and global change on coral reefs', *PLoS Biology*, vol 6, no 2, e54

McClanahan, T. R., Graham, N. A. J., Calnan, J. M. and MacNeil, M. A. (2007) 'Toward pristine biomass: Reef fish recovery in coral reef marine protected areas in Kenya', *Ecological Applications*, vol 17, no 4, pp1055–1067

Moore, F. and Best, B. (2001) 'Coral reef crisis: Causes and consequences', in B. Best and A. Bornbusch (eds) *Global Trade and Consumer Choices: Coral Reefs in Crisis*, American Association for the Advancement of Science, New York, pp5–9

Nyström, M., Folke, C. and Moberg, F. (2000) 'Coral reef disturbance and resilience in a human-dominated environment', *Trends in Ecology and Evolution*, vol 15, no 10, pp413–417

Nyström, M., Graham, N. A. J., Lokrantz, J. and Norström, A. V. (2008) 'Capturing the cornerstones of coral reef resilience: Linking theory to practice', *Coral Reefs*, vol 27, no 4, pp795–809

Paine, R. T. (1966) 'Food web complexity and species diversity', *American Naturalist*, vol 100, no 910, pp65–75

Pandolfi, J. M. (1999) 'Response of Pleistocene coral reefs to environmental change over long temporal scales', *American Zoologist*, vol 39, no 1, pp113–130

Pandolfi, J. M., Bradbury, R. H., Sala, E., Hughes, T. P., Bjorndal, K. A., Cooke, R. G., McArdle, D., McClenachan, L., Newman, M. J. H., Paredes, G., Warner, R. R. and Jackson, J. B. C. (2003) 'Global trajectories of the long-term decline of coral reef ecosystems', *Science*, vol 301, no 5635, pp955–958

Precht, W. F. and Aronson, R. B. (2004) 'Climate flickers and range shifts of reef corals', *Frontiers in Ecology and the Environment*, vol 2, no 6, pp307–314

Reaka-Kudla, M. L. (1997) 'The global diversity of coral reefs: A comparison with rainforests', in M. L. Reaka-Kudla, D. E. Wilson and E. O. Wilson (eds) *Biodiversity II: Understanding and Protecting our Biological Resources*, Joseph Henry Press, Washington, DC, pp83–108

Roberts, J. M., Wheeler, A. J. and Freiwald, A. (2006) 'Reefs of the deep: The biology and geology of cold-water coral ecosystems', *Science*, vol 312, no 5773, pp543–547

Selig, E. R. and Bruno, J. F. (2010) 'A global analysis of the effectiveness of marine protected areas in preventing coral loss', *PLoS ONE,* vol 5, no 2, e9278

Silverman, J., Lazar, B., Cao, L., Caldeira, K. and Erez, J. (2009) 'Coral reefs may start dissolving when atmospheric CO_2 doubles', *Geophysical Research Letters*, vol 36, L05606

Spalding, M. D., Ravilious, C. and Green, E. P. (2001) *World Atlas of Coral Reefs*, University of California Press, Berkeley

Tager, D., Webster, J. M., Potts, D. C., Renema, W., Braga, J. C. and Pandolfi, J. M. (2010) 'Community dynamics of Pleistocene coral reefs during alternative climatic regimes', *Ecology*, vol 91, no 1, pp191–200

Veron, J. E. N., Hoegh-Guldberg, O., Lenton, T. M., Lough, J. M., Obura, D. O., Pearce-Kelly, P., Sheppard, C. R. C., Spalding, M., Stafford-Smith, M. G. and Rogers, A. D. (2009) 'The coral reef crisis: The critical importance of <350 ppm CO_2', *Marine Pollution Bulletin*, vol 58, no 10, pp1428–1436

Wilkinson, C. (2008) *Status of Coral Reefs of the World: 2008*, Global Coral Reef Monitoring Network and Reef and Rainforest Research Centre, Townsville, Australia

6

Preserving Life: Halting Marine Biodiversity Loss and Establishing Networks of Marine Protected Areas in 2010 and Beyond

Marjo Vierros, Biliana Cicin-Sain, Salvatore Arico and Christophe Lefebvre

Background: The Importance of 2010

The United Nations (UN) declared 2010 the International Year of Biodiversity in celebration of life on Earth and the value of biodiversity for our lives. It was also a milestone year for the Convention on Biological Diversity (CBD): the tenth meeting of the Conference of the Parties to the CBD (CBD COP-10) took place in Nagoya, Japan, in October 2010, and considered what progress had been made towards the 2010 biodiversity target. This target, adopted in 2002, committed the parties to the CBD to achieve a significant reduction in the current rate of biodiversity loss at the global, regional and national level by 2010, as a contribution to poverty alleviation and to the benefit of all life on Earth. This target was later endorsed by the World Summit on Sustainable Development (WSSD) as well as the UN General Assembly.

In 2004, at the seventh meeting of the CBD Conference of the Parties (COP), the parties to the CBD adopted a number of sub-targets to clarify the 2010 biodiversity target, and to provide a flexible framework upon which national and/or regional targets may be developed. At the eighth meeting of the COP in 2006, these sub-targets were applied to various biomes, including the marine environment. These targets call for the effective conservation of at least 10 per cent of each of the world's marine and coastal ecological regions; and for the effective protection of particularly vulnerable marine habitats, such as tropical and cold-water coral reefs, seamounts, hydrothermal vents, mangroves, seagrasses, spawning grounds and other vulnerable marine areas.

Fifteen years have now passed since the CBD parties drafted the Jakarta Mandate on Marine and Coastal Biological Diversity. The Jakarta Mandate, which originated from a Ministerial Statement at the second meeting of the CBD COP in Jakarta, Indonesia, in 1995, referred to a new global consensus on the importance of marine and coastal biodiversity. The ministerial statement reaffirmed the critical need for the COP to address the conservation and sustainable use of marine and coastal biological diversity, and urged parties to initiate immediate action to implement COP decisions on this issue. The Jakarta Mandate was operationalized through a programme of work on marine and coastal biodiversity in 1998, which was reviewed and updated by the seventh meeting of the COP in 2004, and was due to be reviewed again in 2010. Centred on the principles of the ecosystem approach and the precautionary approach, the programme of work provides a set of activities for countries to implement according to their national priorities. The activities are grouped under five programme elements that were seen to be global priorities: implementation of integrated marine and coastal area management; sustainable use of marine and coastal living resources; marine and coastal protected areas; mariculture; and invasive alien species (IAS).

The above milestones for the CBD on marine and coastal biodiversity guaranteed there was considerable focus on oceans and coasts at the CBD COP-10 in Nagoya. Additional urgency for further activities aimed towards the protection of the marine environment came from the targets of the WSSD. In particular, we are only a year away from a target agreed to by the WSSD to develop and facilitate the use of diverse approaches and tools, including the ecosystem approach, the elimination of destructive fishing practices, the establishment of marine protected areas (MPAs) consistent with international law and based on scientific information, including representative networks by 2012 and time/area closures for the protection of nursery grounds and periods, proper coastal land use, watershed planning and the integration of marine and coastal areas management into key sectors. This target recognized the important role representative networks of MPAs have in protecting samples of all biodiversity found in the world's oceans so as to ensure their health and survival for the benefit of present and future generations. The target also recognized that, while MPAs are important and have been proven to be successful in reaching biodiversity and fisheries goals, there also exist other tools that can be applied, alone or in combination, to bring about notable benefits to biodiversity and people.

As the deadlines to meet these targets draw near or have already passed, the year 2010 will be remembered as a time of taking stock of why we were unable to reach the 2010 biodiversity target, and what more could be done to reduce the rate of biodiversity loss in the future. We also need to contemplate priority activities needed to reach the 2012 target in regards to networks of MPAs. Our understanding of biodiversity, though far from complete, has improved since the inception of the CBD's Jakarta Mandate and the subsequent programme of work on marine and coastal biodiversity. New drivers of biodiversity loss have appeared, with climate change likely to cause increasing impacts in the future. Thus, it is time to re-evaluate our approach towards conservation and sustainable use of marine biodiversity, and to agree on actions that are most likely to bring about resilient ecosystems and species that can withstand a changing climate and continue to provide the goods and services on which people depend.

The Case for Conserving Marine and Coastal Biodiversity

The case for conserving marine and coastal biodiversity is a compelling one. Biodiversity in the oceans and coastal areas provides numerous benefits to people, including: food resources; regulation of the Earth's climate; and cancer-curing medicines. According to calculations made by The Economics of Ecosystems and Biodiversity (TEEB) project, the value of coral reefs to humankind is between US$130,000 and US$1.2 million per hectare (ha), per year (*Science Daily*, 2009). Mangroves provide an estimated benefit of US$584/ha for local communities due to collected wood and non-wood forest products, US$987/ha for providing nursery for offshore fisheries and US$10,821/ha for coastal protection against storms, totalling US$12,392/ha. This figure does not take into consideration other services, such as carbon sequestration, provided by mangroves. Regardless, the figure is an order of magnitude larger than the benefits of converting the mangroves to shrimp farming (Hanley and Barbier, 2009). The services seagrasses provide in the form of nutrient cycling are valued at an estimated US$1.9 trillion per year, while their support for commercial fisheries is estimated to be worth as much as US$3500/ha per year (Waycott et al, 2009).

The UN Food and Agriculture Organization (FAO) estimates that fish provide more than 2.6 billion people with at least 20 per cent of their animal protein intake. This figure includes protein from a total of more than 1000 species that are harvested from the world's capture fisheries (FAO, 2007b).

As we start to better understand the role of biodiversity in maintaining the Earth's climate regulating system, the case for biodiversity conservation becomes even more urgent. It is estimated that approximately 93 per cent of the Earth's carbon dioxide is stored and cycled through the oceans and that approximately 50 per cent of the carbon in the atmosphere that becomes bound or 'sequestered' in natural systems is cycled into the seas and oceans (Nellemann et al, 2008). An estimated 55 per cent of all carbon in living organisms is stored in mangroves, marshes, seagrasses, coral reefs and macro-algae (TEEB, 2009) making the decline in many of these ecosystems (see section below) even more of a concern.

Status and Trends in Marine and Coastal Biodiversity

Despite the demonstrated economic and social values provided by marine biodiversity, it is evident from the best available scientific information that the 2010 biodiversity target was not reached for oceans and coasts globally. The global decline in marine biodiversity has been well documented. Available indicators – such as the Marine Living Planet Index, which tracks population trends of representative marine species – show a continued decline overall in the abundance, diversity and distribution of marine species.

According to available information from ecosystems ranging from coastal estuaries and shellfish reefs to deep-sea seamounts and pelagic fisheries, biodiversity in the oceans is declining, as demonstrated by the statistics below:

- According to the Global Coral Reef Monitoring Network (GCRMN), we have effectively lost 19 per cent of the original area of coral reefs; 15 per cent are seriously threatened with loss within the next 10–20 years; and an additional 20 per cent are under threat of loss in 20–40 years (Wilkinson, 2008).
- Oyster reefs have declined more than 90 per cent from their historical levels, making them one of the most imperiled marine habitats on Earth.
- Wetlands and seagrass communities continue to decline worldwide, drastically reducing their ability to provide valuable services in supporting fisheries, carbon sequestration and protecting coastal areas from storms. The rate of seagrass disappearance has been estimated to be $110km^2/yr$ since 1980, with 29 per cent of the known areal extent now lost since 1879. The rate of loss is accelerating (Waycott et al, 2009).
- Many cold water coral reefs have been damaged by bottom-fishing activities. While the extent of this damage has not been quantified, most reefs studied thus far show physical damage from trawling activities. In addition, these reefs are especially threatened by ocean acidification, estimates predict that 70 per cent of the 410 known locations with deep-sea corals may be in aragonite-under saturated waters by 2099 (Secretariat of the CBD, 2008).
- Fisheries stocks assessed since 1977 have experienced an 11 per cent decline in total biomass globally, with considerable regional variation (Worm et al, 2009).
- Globally, dead zones (oxygen deficient zones) are increasing due to nutrient overenrichment from marine pollutants, as is the spread of invasive alien species (Nellemann et al, 2008).

While the picture remains grim, there are also some bright spots and progress was made towards the achievement of the 2010 biodiversity target in relation to certain species and ecosystems. For example, available data indicates that the net loss of mangroves, while still very high, may have slowed down (from 185,000ha/yr loss in the 1980s to 102,000ha/yr during 2000–2005), possibly due to massive replanting campaigns following growing attention to the value of mangroves in the wake of the 2004 tsunami (FAO, 2007a). While the health of coral reefs near major population centres show a continued decline, reefs in the Indian Ocean and Western Pacific have shown significant recovery since the devastating 1998 bleaching events (Wilkinson, 2008). There are also many examples of local success stories, where drivers of biodiversity loss have been successfully addressed and resources have recovered due to protection measures. While these efforts should be celebrated and lessons learned, collectively they are not enough to slow the loss of biodiversity globally.

Given these declining trends, an important task now rests with the global oceans community to assess both the global status of marine biodiversity and progress made in achievement of biodiversity targets, as well as to outline the next steps in moving forward on the biodiversity agenda in upcoming years.

The Implications of Climate Change for Marine and Coastal Biodiversity

The impacts of climate change, which are predicted to increase in the future, have significant implications for marine biodiversity and will serve to exacerbate the negative impacts of other harmful human activities. As biodiversity is essential to ecosystem functions, even slight impacts on marine biodiversity can have severe implications for global ecosystems.

Increases in water temperature will cause more frequent and severe coral bleaching events. Mass bleaching is expected to take place on an annual basis in the future, departing from the 4–7 years return-time of El Niño events. Coral bleaching will be exacerbated by the effects of degraded water quality and increases in the frequency and severity of extreme weather events (Veron et al, 2009).

Ocean acidification will become a serious problem, reducing the biocalcification of tropical and cold-water coral reefs, as well as other shell-forming organisms, such as calcareous phytoplankton. This will affect the entire marine food chain and result in less diverse biological communities (Secretariat of the CBD, 2009). According to the Inter-Academy Panel Statement on Ocean Acidification (IAP) in June 2009, if current emission rates continue, models suggest that all coral reefs and polar ecosystems will be severely affected by 2050 or potentially even earlier. Limiting atmospheric carbon dioxide levels significantly below 350ppm will likely ensure the long-term viability of coral reefs (Coral Reef Crisis Working Group Meeting, 2009).

Rising ocean temperatures and increases in freshwater input from the melting of polar ice formations are likely to adversely impact ocean circulation, including potentially reducing the intensity and frequency of large-scale water exchange mechanisms. This, in turn, would impact both nutrient and larval transport systems and increase the risk of oxygen deficient zones (Policy Brief on Climate, Oceans and Security, 2010). Biodiversity in the deep oceans could also be affected as warming oceans may result in large variations in the amount of organic material reaching the seafloor (Smith et al, 2009a). Climate change will reduce the human benefits derived from marine biodiversity. Climate change, and its impacts on marine biodiversity, has significant implications for food security. A large portion of the world's population is heavily dependent upon ocean resources for sustenance. Impacts on marine food supplies will likely serve to exacerbate worldwide hunger and may lead to resource conflicts in certain areas.

Some climate change response strategies may also adversely impact marine biodiversity. For example, significant concern has been expressed regarding the potential impacts of large-scale ocean fertilization on marine species, habitats and ecosystem functions. As a result, the CBD has called for a precautionary approach to ensure that ocean fertilization activities do not take place until there is an adequate scientific basis on which to justify such activities (Secretariat of the CBD, 2009).

Effective management and protection of marine areas, including through MPAs, will enhance the resilience of biodiversity to the impacts of climate change by removing other external stress factors, and thus providing a better opportunity for adaptation (Smith et al, 2009b).

Implementation of Networks of MPAs and Other Measures for Conservation and Sustainable Use of Biodiversity

At the present time, only 0.5 per cent of the oceans overall are covered by MPAs. More progress has been made closer to shore, with 6.3 per cent of territorial sea now protected, an increase from 2.9 per cent in 1990 and 5 per cent in 2000 (Statistics courtesy of United Nations Environment Programme World Conservation Monitoring Centre (UNEP-WCMC) personal communication). While this figure falls short of the 10 per cent target set by the CBD, it still demonstrates that considerable national action towards the conservation of the marine environment has been undertaken by countries individually or collectively. Figures for national Exclusive Economic Zones (EEZs) have not been calculated as of yet, given difficulties posed to such calculations by ongoing extended continental shelf claims. These statistics also demonstrate that deep-sea and open ocean areas beyond national jurisdiction remain some the most underprotected regions on Earth.

According to national reports submitted to the CBD, almost all countries now have one or more MPA and many have established national networks of MPAs. Recently, the establishment of spatially expansive MPAs, such as the Phoenix Islands Protected Area, in Kiribati; the Papah naumoku kea Marine National Monument in the Northwestern Hawaiian Islands; and the Chagos Islands MPA (set up by the UK government) have greatly increased the amount of protected areas in the ocean. Ambitious regional initiatives, such as the Micronesia Challenge, the Caribbean Challenge and the Coral Triangle Initiative are also set to protect important marine biodiversity and demonstrate a positive trend in the use of MPAs to protect marine biodiversity and sensitive ecosystems. At the global level, the World Network of Biosphere Reserves under the United Nations Educational, Scientific and Cultural Organization's (UNESCO) Man and the Biosphere Programme counts more than 150 coastal marine sites. With a rise in marine spatial planning and large-scale bioregional classification initiatives, many countries are developing MPA networks as part of comprehensive management regimes, thus implementing MPAs in a broader ecosystem approach context.

Progress is also being made at the regional level, especially through the work of various Regional Seas Programmes, including those of the UNEP. In the North Atlantic, for example, the Oslo/Paris Convention for the Protection of the Marine Environment of the North-East Atlantic (OSPAR) Commission is working to develop an ecologically coherent network of MPAs by the end of 2010. The regional approach to protection of the marine environment, which is emphasized in the United Nations Convention on the Law of the Sea (UNCLOS), can often prove to be the most appropriate scale to encourage inter-sectoral cooperation in the protection of the marine environment and to move forward in the implementation of networks of marine protected areas.

Despite notable progress being made, the global MPA network is not yet representative of all biodiversity. Of the near-shore habitats, coral reefs and mangroves are relatively well protected, while seagrasses and shellfish reefs are afforded relatively less protection in existing MPA systems. Very few spawning aggregations are protected. Approximately 43 per cent of all MPAs (or about 65 per cent of the total area that is protected) lie in the tropics (between 30°N and 30°S), with most of the remainder in the northern hemisphere. Intermediate latitudes (20°N–50°N) and the southern temperate and polar latitudes are least represented (UNEP-WCMC, 2008).

Deep-sea and open ocean habitats are also afforded very little protection, particularly in marine areas beyond the limits of national jurisdiction. Regional and national initiatives such as the OSPAR network of MPAs have begun to identify and, in some cases, declare areas for protection. Within national jurisdiction, some countries are now actively seeking to protect deeper water habitats within their EEZs. Deep-sea pelagic habitats are presently afforded almost no protection.

There is no comprehensive information available regarding the management effectiveness of MPAs globally, although some national studies exist. According to anecdotal evidence, the management of many MPAs is still lacking. Some studies also show that MPAs have been more effective in reaching ecological rather than social goals (Christie, 2004).

Challenges

Marine and coastal biodiversity loss is caused by multiple drivers that are intensifying. These drivers include development and land-use patterns, pollution, unsustainable fishing, IAS and other impacts. Coastal populations are predicted to increase, with 50 per cent of the world's population expected to live along the coasts by 2015. Projections from UNEP estimate that as much as 91 per cent of all temperate and tropical coasts will be heavily impacted from increased levels of development by 2050 (Nellemann et al, 2008). The impacts of climate change are also predicted to increase in the future, thereby exacerbating effects on marine biodiversity.

The drivers of biodiversity loss cannot be controlled by environmental agencies alone, and mainstreaming of biodiversity concerns into the activities of other sectors is often lacking. Slowing biodiversity loss requires the involvement of all sectors, including fisheries, forestry, agriculture, coastal development and shipping. The best means to reverse the decline of biodiversity is through all ocean users incorporating biodiversity-relevant priorities into their activities.

The economic and social benefits and values of marine and coastal biodiversity are often not well understood by decision-makers, resulting in limited political will to undertake action towards biodiversity protection. This is particularly true if the required action is likely to be unpopular in the short term, such as limiting development and extractive or other revenue-generating activities. The benefits of protection will only be apparent long after national election cycles have passed. The lack of appreciation regarding the goods and services provided by marine biodiversity may also explain the limited application of the precautionary approach in management.

The economic and social costs and benefits of biodiversity conservation are not equitably shared. The short-term costs of, for example, establishing an MPA, may be disproportionately borne by certain communities or resource users, while benefits may be shared by a larger group of users and could take a significant amount of time to materialize. In many developing countries, biodiversity conservation may be too costly when compared to other more immediate needs. Certain research activities, which can lead to improvements in scientific knowledge and provide a stronger basis for conservation efforts, can prove to be beyond the financial and technical capabilities of many developing nations.

Conservation measures do not always respect local cultural norms and social structures, and may not bring direct benefits to communities. In some parts of the world, conservation efforts have often ignored local and traditional knowledge and land/sea tenure systems in favour of a top-down scientific model, resulting in social and cultural losses to coastal communities. In many cases, MPAs have been more successful in bringing ecological rather than social benefits, and have thus failed to gain the support of communities.

Available data and information relating to the marine environment is not always easily accessible or well-organized and new research and monitoring efforts are not comprehensive or responsive to management needs. There is currently no comprehensive global assessment/monitoring of the status of biodiversity in the oceans, robust indicators are lacking and existing efforts are not always well-coordinated. As a result we still have very little understanding of what we have and what we stand to lose.

The Way Forward: General Recommendations for Future Action

Recommended actions[1]

- Ensure that the conservation and sustainable use of marine biodiversity becomes the common concern of every country by creating an improved understanding of its economic and non-economic values. In many cases, the goods and services provided by biodiversity are not well understood, and therefore not highly valued in national policies that seek to maximize economic development. Thus, it will be important to demonstrate the role biodiversity in the oceans plays in supporting human life and livelihoods through promoting its economic, social and cultural values. While economic valuation activities have recently become more common and have been effectively used in support of conservation measures, they have generally focused on selected ecosystems, such as coral reefs. Much less is known, for example, about the economic values of deep-sea ecosystems. Some studies have calculated the costs of specific conservation action (such as the cost to establish a global network of MPAs), but increased focus should also be paid to calculating the economic costs of inaction (failing to undertake conservation measures) in the long term. The work of initiatives such as TEEB project should be supported and their results widely disseminated to decision-makers.

- Ensure that marine biodiversity concerns are mainstreamed into the work of all relevant sectors and that all stakeholders are included in the visioning, planning and management processes, in the overall framework of the ecosystem approach.
- Mainstreaming and integration can be supported through national initiatives such as marine spatial planning, where all ocean users are involved in the planning and management process and work towards a common goal. Mainstreaming has been most successful in countries where biodiversity is self-evidently a crucial component of national wealth (for example, in the form of tourism income), and thus each stakeholder has an incentive to participate in the development and implementation of marine spatial planning. Communication and participatory approaches, such as multi-stakeholder dialogues are an important tool in ensuring mainstreaming of biodiversity into all relevant sectoral policies.
- Broadly implement ecosystem-based management, including through the establishment of networks of MPAs. There is an urgent need to improve protection and management of the oceans, particularly in areas that are currently under-represented in MPA systems. Thus, there is a need to increase the coverage of MPAs and ensure that MPA networks are representative of the full range of biodiversity in the oceans, including deep seas and pelagic areas. Attention should be paid to ensuring that MPAs are well-managed and that they provide both ecological and socio-economic benefits. MPAs alone are not enough, and the areas outside them (whether land or sea) need to be sustainably managed as well, keeping in mind that improved management will increase the resilience of marine and coastal biodiversity to the impacts of climate change. Ultimately, MPAs are central to maintaining the processes upon which the sustainability of the marine environment depends. To this end, they need to be complemented by other actions including the establishment of ecological corridors. By integrating MPAs into

marine spatial planning efforts, due consideration will be given to specific stakeholder needs, in the context of an integrated approach to management.

- Create a global list of marine areas of ecological and biological significance, also taking into account representativity. Significant progress is being made towards the identification of marine areas in need of special protection. For example, the work of the CBD in biogeographic classification and criteria for the identification of ecologically and biologically sensitive areas (European Biosafety Association, EBSA) (CBD Decision IX/20, Annex1) outlines an important step toward improvements in marine protection capability in deep and open oceans. The efforts should be further supported and implemented through the development of a coherent global list of marine areas in need of special protection, beyond the limits of national jurisdiction, which will serve to provide scientific validation, and ultimately facilitate the creation of MPA networks at the national, regional and global levels, in the context of the law of the sea, and under appropriate legal instruments, which may include regional arrangements, such as Regional Fishery Management Organizations, Regional Seas, or appropriate global instruments. Lessons can be learned from successful regional efforts, including those under OSPAR and the Convention on the Conservation of Antarctic Marine Living Resources (CCAMLR).
- Strengthen capacity for the creation and management of networks of MPAs. As capacity is a central factor in the effective management of marine resources, especially in developing countries and Small Island Developing States (SIDS), capacity-building is essential to the creation and management of networks of MPAs. International solidarity initiatives could support the strengthening of capacity and the development of national agencies dedicated to the management of MPAs, in the context of integrated coastal and ocean management. Concrete actions aimed at strengthening capacity for MPAs should be complemented by capacity-building activities in related areas, namely, sustainable fishing, marine scientific research and the management of adverse impacts that human activities have on the marine environment.
- Encourage the development of an institutional and inter-sectoral approach to marine protection in areas beyond national jurisdiction. In light of increased activity in marine areas beyond the limits of national jurisdiction – including deep-sea fishing and oil and gas drilling, as well as new and emerging activities such as bioprospecting – the international community, and especially the UN, should seek to address existing legal and regulatory gaps in the international framework for protection of biodiversity in these areas. Options for the development of legal institutional mechanisms as well as methods to better utilize existing management frameworks should be developed and implemented to improve protection of marine biodiversity in areas beyond national jurisdiction.

- Support and learn from creative local solutions. There are many ways to improve the management of marine biodiversity, and the best solutions are often local. These bottom-up approaches are embedded in local knowledge and have a unique social and cultural context. Sharing information about local success stories may provide new ideas for management worldwide.
- Improve the informational basis for management and for assessing progress by ensuring that management action is based on the best available scientific information. Existing information should be made more accessible to all users and new targeted research, monitoring and data analysis activities supported. Initiatives such as the Regular Process for Global Reporting and Assessment of the State of Oceans, the continuation of the Census of Marine Life, the Global Ocean Biodiversity Initiative (GOBI) and the Intergovernmental Science-Policy Platform on Biodiversity and Ecosystem Services (IPBES) have much to offer for improving protection and management of biodiversity, and should be supported and coordinated. The role of local knowledge in management should also be recognized and efforts to integrate it with scientific knowledge supported.
- Address capacity and equity. Marine biodiversity protection efforts should squarely address issues of capacity and equity. The concept of capacity is broad and includes the creation of an enabling policy and legislative environment; the availability of adequate staff and resources; the acquisition of specific skills through training and other means; and the availability of necessary equipment, such as hardware and software, including related appropriate technology transfer. Capacity needs vary greatly between countries and regions, and successful capacity-building must be closely targeted to meet the needs of the recipients. Issues of equity in access and benefit sharing regarding marine genetic resources are a salient concern of developing countries and must also be directly addressed.
- Improve long-term financing. A majority of projects relating to conservation and sustainable use of marine and coastal biodiversity suffer from limited financial support, preventing them from becoming sustainable in the long term. There is a need for improved and forward-looking financial planning to allow networks of MPAs and other management efforts to become self-sustaining. Funding will need to fit the scope of the projects and the needs of the recipient countries or organizations, and may include small or larger grants, financing from the Global Environment Facility (GEF) and other donors, user fees, conservation trusts and other means. The development of business planning skills may be a fundamental component of project planning. In addition to management, improved funding is required for research, monitoring and enforcement. It should be noted that where a management activity has community support, enforcement costs are likely to be reduced.

Opportunities to advance the marine biodiversity agenda in the future

The CBD Parties adopted an extensive decision on marine and coastal biodiversity at their 10th meeting (decision X/29). This decision covered a number of important topics that were considered to be future priorities. Included in the decision were sections on identification of ecologically and biologically significant areas (EBSAs); scientific and technical aspects relevant to environmental impact assessment in marine areas; impacts of ocean acidification; and impacts of unsustainable fishing, and illegal, unreported and unregulated (IUU) fishing, ocean fertilization, underwater noise, and other human activities. The adopted decision provides for an ambitious programme of future work, addressing issues ranging from scientific and technical aspects of EBSAs in marine areas beyond national jurisdiction to the role of biodiversity in climate change mitigation and adaptation. The vital role of tools and approaches such as marine and coastal protected areas and environmental impact assessment continues to be recognized and enhanced.

In addition, the Strategic Plan for Biodiversity 2011–2020 and the Aichi Biodiversity Targets (decision X/2) provide new opportunities for advancing the management and conservation of marine biodiversity. These targets include the following oceans-related components:

- *Target 6*: 'By 2020 all fish and invertebrate stocks and aquatic plants are managed and harvested sustainably, legally and applying ecosystem based approaches, so that overfishing is avoided, recovery plans and measures are in place for all depleted species, fisheries have no significant adverse impacts on threatened species and vulnerable ecosystems and the impacts of fisheries on stocks, species and ecosystems are within safe ecological limits.'
- *Target 10*: 'By 2015, the multiple anthropogenic pressures on coral reefs, and other vulnerable ecosystems impacted by climate change or ocean acidification are minimized, so as to maintain their integrity and functioning.'
- *Target 11*: 'By 2020, at least 17 per cent of terrestrial and inland water areas, and 10 per cent of coastal and marine areas, especially areas of particular importance for biodiversity and ecosystem services, are conserved through effectively and equitably managed, ecologically representative and well-connected systems of protected areas and other effective area-based conservation measures, and integrated into the wider landscapes and seascapes.'

Another opportunity for future work will come in the context of the Rio+20 process. Rio+20 will provide a framework to develop long-term goals, targets and specific strategies related to marine biodiversity in the context of the international debate on sustainable development.

The United Nations Framework Convention on Climate Change (UNFCCC) and other climate-relevant processes can also assist in developing a comprehensive programme on all aspects of oceans and climate within and outside the UNFCCC, with the participation of the Intergovernmental Panel on Climate Change (IPCC).

Other processes, including but not limited to the UN General Assembly process on marine biodiversity beyond the limits of national jurisdiction, the FAO work on fisheries and the Regular Process for Global Reporting and Assessment of the State of Oceans all contribute towards the attainment of these targets. On the regional level, Regional Seas programmes and Regional Fisheries Management Organizations continue to advance the biodiversity agenda.

The Vision for the Future: The Nagoya Oceans Statement

The Oceans Day at Nagoya, organized by the Global Forum on Oceans, Coasts and Islands and partners took place on 23 October 2010. The intention was to raise further awareness of the threats to – and the importance of – marine biodiversity, and the need to ensure relevant, timely and coordinated international action (Vierros and Appiott 2011). The Oceans Day resulted in the Nagoya Oceans Statement, which is intended to provide input to high level discussions related to marine and coastal biodiversity, both within the CBD and other international fora. The priority actions in the Statement will also help efforts to make progress towards the Aichi Biodiversity Targets for oceans and coasts.

The Nagoya Oceans Statement reaffirms the long-term vision put forward in the CBD Elaborated Programme of Work on Marine and Coastal Biological Diversity (decision VII/5, Annex 1) 'to halt the loss of marine and coastal biological diversity nationally, regionally and globally, restore degraded marine habitats, and secure the capacity of the oceans to continue to provide goods and services' and supports subsequent updates and decisions related to its implementation.

In accordance with decisions IV/5 and VII/5, the Nagoya Oceans Statement recognizes that all activities undertaken towards their implementation shall be guided by:

- The ecosystem approach
- The precautionary approach
- The best available science, including the scientific, technical and technological knowledge of local and indigenous communities

The Nagoya Oceans Statement reaffirms the Jakarta Mandate, which referred to a new global consensus on the importance of marine and coastal biodiversity. However, recognizing that 15 years have passed since the adoption of the Jakarta Mandate, and that new drivers of biodiversity loss and new scientific research have led the global community to further develop and refine approaches to the conservation and sustainable use of marine and coastal biodiversity, it is essential to highlight the importance of:

- Accelerating efforts to establish representative networks of MCPAs consistent with international law and based on scientific information, with particular attention to addressing gaps in the current global system of MCPAs, including for deep-sea and open-ocean areas, taking into account lessons learned from establishment of MCPAs in various countries, including the creation and/or designation of dedicated entities charged to carry out this function;
- Undertaking the establishment of MCPAs in the broader context of marine spatial planning and integrated marine and coastal area management (IMCAM) at various levels, which will provide for the integration of biodiversity objectives into sectoral and development programmes;
- Developing regional transboundary cooperation, building on existing regional bodies, for example, regional seas agreements, in the context of the ecosystem approach, for establishing marine conservation connectivity within coherent and resilient MCPA networks;
- Addressing cumulative impacts of human activities on the marine environment through the application of environmental impact assessment (EIA) and strategic environmental assessment (SEA);
- Developing and implementing priority actions that enhance the contribution of marine and coastal ecosystems to climate change mitigation and adaptation, including maintaining the capacity of oceans to store carbon and strengthening the resilience of marine and coastal systems to the impacts of climate change and ocean acidification;
- Recognizing the special importance of marine biodiversity in SIDS and developing nations, and the need to build capacity to address threats to biodiversity in these areas;
- Improving engagement of all ocean stakeholders, including indigenous and local communities, in the conservation and sustainable use of biodiversity, including through capacity-building and the integration of socio-economic and cultural considerations and traditional knowledge into management frameworks;
- Integrating communication, education and public awareness-raising into biodiversity protection strategies to encourage stewardship of marine and coastal biodiversity in current and future generations and, thus, to contribute to meeting the objectives of the Convention on Biological Diversity.

Acknowledgements

The research assistance of Joseph Appiott and Gwenaelle Hamon in the preparation of this chapter is acknowledged with sincere thanks.

Notes

1 This chapter was originally prepared as a Policy Brief on Marine Biodiversity and Networks of Marine Protected Areas to elicit discussion and debate at the Fifth Global Oceans Conference 2010 (3–7 May 2010 at UNESCO, Paris), and subsequently revised by the authors. Special thanks are due to the French Marine Protected Areas Agency and to the Global Environment Facility for their support related to tracking progress on the implementation of global commitments on oceans emanating from the 2002 World Summit on Sustainable Development. Any opinions expressed in this chapter are solely the opinions of the contributing authors and do not reflect institutional positions on the part of the Global Forum on Oceans, Coasts, and Islands nor of any of the sponsoring organizations of the Global Oceans Conference 2010.

References

Christie, P. (2004) 'Marine protected areas as biological successes and social failures in Southeast Asia', *American Fisheries Society Symposium*, vol 42, pp155–164

Coral Reef Crisis Working Group Meeting (2009) 'Statement of the Coral Reef Crisis Working Group Meeting, The Royal Society 6 July 2009', Royal Society of London, http://static.zsl.org/files/statement-of-the-coral-reef-crisis-working-group-890.pdf

FAO (Food and Agriculture Organization) (2007a) 'The world's mangroves 1980–2005', FAO Forestry Paper no 153, Food and Agriculture Organization, Rome

FAO (2007b) *The World's Aquatic Genetic Resources: Status and Needs*, background document CGRFA-11/07/15.2 for the Eleventh Regular Session of the Commission on Genetic Resources for Food and Agriculture, ftp://ftp.fao.org/docrep/fao/meeting/014/j9581e.pdf

Hanley, N. and Barbier, E. B. (2009) *Pricing Nature: Cost-Benefit Analysis and Environmental Policy*, Edward Elgar, London

Nellemann, C., Hain, S. and Alder, J. (eds) (2008) *In Dead Water: Merging of Climate Change with Pollution, Over-harvest, and Infestations in the World's Fishing Grounds*, United Nations Environment Programme, GRID-Arendal, Norway

Policy Brief on Climate, Oceans and Security, 'Fourth Global Conference on Oceans, Coasts and Islands: Working Group on Climate, Oceans, and Security', www.globaloceans.org/globaloceans/sites/udel.edu.globaloceans/files/Climate-and-Oceans-PB-April2.pdf

Science Daily (2009) 'What are coral reef services worth? US$130,000 to US$1.2 million per hectare, per year', www.sciencedaily.com/releases/2009/10/091016093913.htm#at

Secretariat of the CBD (Convention on Biological Diversity) (2008) 'Synthesis and review of the best available scientific studies on priority areas for biodiversity conservation in marine areas beyond the limits of national jurisdiction', CBD Technical Series no 37, Convention on Biological Diversity, Montreal

Secretariat of the CBD (2009) 'Scientific synthesis of the impacts of ocean fertilization on marine biodiversity', CBD Technical Series no 45, Convention on Biological Diversity, Montreal, www.cbd.int/doc/publications/cbd-ts-45-en.pdf

Smith, K. L. Jr, Ruhl, H. A., Bett, B. J., Billett, D. S. M., Lampitt, R. S. and Kaufmann, R. S. (2009a) 'Climate, carbon cycling, and deep-ocean ecosystems', *PNAS*, vol 106, no 46, pp19211–19218

Smith, S. E., Meliane, I., White, A., Cicin-Sain, B., Snyder, C. and Danovaro, R. (2009b) 'Climate change on marine biodiversity and the role of networks of marine protected areas', in *Oceans and Climate Change: Issues and Recommendations for Policymakers and for the Climate Negotiations*, policy briefs prepared for the World Ocean Conference, 11–15 May 2009, Manado, Indonesia, pp131–136

TEEB (The Economics of Ecosystems and Biodiversity) (2009) 'Climate issues update', www.teebweb.org

UNEP-WCMC (United Nations Environment Programme World Conservation Monitoring Centre) (2008) *State of the World's Protected Areas: An Annual Review of Global Conservation Progress*, United Nations Environment Programme World Conservation Monitoring Centre, Cambridge

Veron, J. E. N., Hoegh-Guldberg, O., Lenton, T. M., Lough, J. M., Obura, D. O., Pearce-Kelly, P., Sheppard, C. R. C., Spalding, M., Stafford-Smith, M. G. and Rogers, A. D. (2009) 'The coral reef crisis: The critical importance of <350 ppm CO_2', *Marine Pollution Bulletin*, vol 58, no 10, pp1428–1436

Vierros, M. and Appiott, J. (2011) Summary of the Oceans Day at Nagoya, held at the 10th Meeting of the Conference of the Parties to the Convention on Biological Diversity (CBD COP-10), 23 October 2010, Nagoya, Japan, Global Forum on Oceans, Coasts, and Islands, available at http://globaloceans.org/sites/udel.edu.globaloceans/files/Nagoya_Summary.pdf

Waycott, M., Duarte, C. M., Carruthers, T. J. B., Orth, R. J., Dennison, W. C., Olyarnik, S., Calladine, A., Fourqurean, J. W., Heck, K. L. Jr, Hughes, A. R., Kendrick, G. A., Kenworthy, W. J., Short, F. T., Williams, S. L. (2009) 'Accelerating loss of seagrasses across the globe threatens coastal ecosystems', *PNAS*, vol 106, no 30, pp12377–12381

Wilkinson, C. (2008) *Status of Coral Reefs of the World: 2008*, Global Coral Reef Monitoring Network and Reef and Rainforest Research Centre, Townsville, Australia

Worm, B., Hilborn, R., Baum, J. K., Branch, T. A., Collie, J. S., Costello, C., Fogarty, M. J., Fulton, E. A., Hutchings, J. A., Jennings, S., Jensen, O. P., Lotze, H. K., Mace, P. M., McClanahan, T. R., Minto, C., Palumbi, S. R., Parma, A. M., Ricard, D., Rosenberg, A. A., Watson, R. and Zeller, D. (2009) 'Rebuilding global fisheries', *Science*, vol 325, no 5940, pp578–585

7

Forests: Thinking and Acting outside the Box

Jan L. McAlpine

Forests offer solutions to our most salient problems: how to fight poverty, how to stabilize the climate, how to establish and maintain livelihoods for people who may otherwise be marginalized, how to protect public health and how to enhance ecosystem services. Forests can help us to achieve the Millennium Development Goals (MDGs). With 192 member states, the United Nations Forum on Forests (UNFF) is the only body other than the United Nations (UN) General Assembly to have universal participation. In UNFF's vision, all countries are forest countries and UNFF's comprehensive strategy is a 360-degree approach.

For historical reasons, however, institutions do not generally mirror this comprehensive approach. Around the world, environmental, social and economic institutions continue to work separately, in silos. As a result, we lose opportunities to solve problems by using all of these institutions to achieve our shared objectives. UNFF is intent on finding ways to bridge these different areas and related institutions, to facilitate and catalyse collaborative work among them. The ability of forests to absorb and store carbon dioxide has recently dominated the world stage, however, without a 360-degree approach that addresses all of the benefits of forests, including sound livelihoods and healthy forest ecosystems, the goal of carbon absorption will not be achieved and cannot be maintained. It is not an option to protect one role of forests without the others. Essential climate mitigation and adaptation will only succeed in the context of a comprehensive approach to forests.

In the past, we made the mistake of valuing forests only for their timber, which led to serious distortions and in some cases a significant decline in forest cover. Policy-makers learned some lessons from this, as reflected in the many resolutions reached by UNFF that underline the critical importance of internalizing the other benefits of forests in their sustainable management. Now, however, we risk going down the same path by valuing forests only for absorbing carbon, even though this absorption is indeed crucial. The funding for forests worldwide is small compared to the amount allocated for Reducing Emissions from Deforestation and Forest Degradation Plus (REDD+) in developing countries, and the role of conservation, sustainable management of forests and enhancement of forest carbon stocks in developing countries has already exceeded a commitment of US$4 billion, which could undermine a comprehensive approach to forests. We must balance and integrate financial support to address all aspects of sustainable forest management.

Wealth of Goods and Services

Forests are of extremely high value for a range of reasons. Firstly, forests are the most important terrestrial reservoir of biodiversity. Time and again scientists have pinpointed them as the most biodiversity-rich ecosystems on Earth and they constitute a wealth of natural heritage not only in themselves, but also for people who use this biodiversity for a variety of purposes, including for water, food and medicinal purposes. To take just one example, a 25-acre fragment of Bornean forest is known to contain more than 700 tree species – a number equal to the tree diversity of the whole of North America.

Secondly, forests are home to a very high level of biomass, hence the immense value of forests in storing carbon. This is especially true of primary forests, both in temperate/boreal and tropical forests, where aged trees can reach exceptional sizes. Likewise, peat forests store carbon not only in living matter, but also accumulate it in the soil where acidic conditions prevent forest debris from decomposing. Today, both these categories of forests are particularly prone to land clearance with primary forests having decreased by more than 40 million hectares since 2000 (FAO, 2010, p5). This has major implications for the role that forests play in global climate change, as is further discussed below.

Forests also constitute the primary resource base for timber and other industries worldwide. The timber market is a global one which millions live on, both in developing and developed countries. In the period 2003–2007, wood removals alone valued over US$100 billion annually. By giving forests direct economic value through selective logging, the timber industry contributes to the maintenance of standing forests. Additionally, although smaller, industries also depend on forests for non-timber forest products and services, such as Brazilian nuts, various oils and ecotourism, among others.

A Safety Net for the Rural Poor

Forests are often located in regions of the world where impoverished rural populations rely crucially on their natural environment as a source of livelihoods. They provide for the daily subsistence needs and essential livelihood services of more than 1.6 billion people. These people directly rely on forests for essential products such as clean water, building materials, food (both animal protein and fruit) and medicine. Many more partially depend on forests for local climatic patterns that ensure sufficient rainfall for nearby crops. Local populations have often been accused of being major drivers of deforestation, but a number of studies have shown that this is not the case.

In light of the multiple values of forests, their daily degradation and disappearance are an immense loss to humanity. At a local level, indigenous and forest-dwelling populations are the ones most affected. The destruction of their only source of livelihoods, and of their environment, is known to lead to economic and social marginalization and impoverishment at the very least. In many cases, these populations have died out altogether as their social fabric falls apart in the absence of forests upon which they crucially depend, both economically and culturally.

Western values, which are increasingly financially based, often fail to capture the importance of forests to both people and the global economy. Many functions that forests fulfil do not have any monetary value and even when they do, they are greatly underestimated. Historically, forests have contributed to the global economy by providing us with products that we take for granted such as cocoa (from the Amazon), coffee (from Ethiopian forests), apples, apricots, walnuts and pears (from Central Asian forests). Yet these facts remain little known and have prevented forests from being recognized for the exceptional wealth that they represent.

Likewise, the great cultural and spiritual importance that forests constitute in the eyes of hundreds of millions is frequently underestimated, if not totally ignored. This value is particularly difficult to quantify, and is thus often ignored or underestimated by some, while it is considered crucial by others. For many people, forests are an integral part of their cultural and social identity, as well as a central component in their systems of beliefs. The importance of forests in their eyes cannot be overestimated.

An Essential Component of Our Lives

Beyond the potentially catastrophic consequences on local populations, on a regional scale both forest degradation and deforestation are known to affect local climate, soil fertility and availability of clean water. Land conversion, especially when it affects rainforests (both temperate and tropical), affects rainfall patterns that nearby rural populations rely on for agriculture. The disappearance of roots to fix the soil quickly leads to soil erosion, especially in tropical countries where soil is thinly spread and easily washed away in the rain. In more dramatic examples, such as Haiti and the Philippines, rapid deforestation in mountainous areas combined with heavy rains have increased the probability of mudslides, which have been known to wipe out entire villages. Greater soil mobility, altered rainfall patterns and the absence of vegetation cover to prevent evaporation also affect water sources, which dry up or get silted.

Together, these factors lead to land degradation and desertification, reduced agricultural output and increasingly difficult access to drinking water. These phenomena usually hit impoverished rural populations first. Consequences include exacerbated poverty at regional scales, famines and large-scale population movements, which in turn lead to unregulated urbanization and social conflict.

Consequences do not stop at the regional level. On a global scale, deforestation affects the world in a variety of ways. Rainforests in particular contribute such a large proportion of freshwater (the Amazon Basin alone providing approximately a fifth of the world's freshwater) that their disappearance will significantly increase freshwater scarcity, leading to greater probability of political conflict over this basic but crucial natural resource. A decrease in the availability of timber could lead to the collapse of entire components of the timber industry, although the current expansion of plantations might mitigate these effects somewhat. Plantations, however, would not solve the irreversible loss of biodiversity that comes with degradation and deforestation.

Regarding the carbon storage function of forests, it is now a well-established fact that forest degradation and deforestation contribute 18–25 per cent of global greenhouse gas (GHG) emissions. Clearing forests releases carbon dioxide into the atmosphere through the burning and decomposition of large amounts of living matter. The fact that Indonesia now ranks third in the world in terms of GHG emissions should not come as a surprise given the country's high deforestation rate. Deforestation has thus become a major driver of global climate change, which is likely to see global temperatures rise and weather patterns change significantly, which in turn will increase poverty, the likelihood of humanitarian crises and possibly social and political conflict.

Understanding Causes of Deforestation

Given the immense value of forests from a variety of perspectives, it might appear puzzling that they continue to be cleared at such alarming rates. The causes of deforestation are complex, numerous and have been the subject of many studies.

In their comprehensive literature review on the subject, Angelsen and Kaimowitz (1999) categorize deforestation causes into 'immediate' and 'underlying'. According to them, the best established immediate cause of deforestation is agricultural prices. One could argue that subsistence farmers would grow fewer crops if they got more money out of them; yet higher prices generally stimulate large-scale, industrialized agriculture. Their point was proven only several years after their publication, when a peak in agricultural prices coincided with a sharp increase in deforestation in the Brazilian Amazon. The same applies for the so-called 'Hamburger Connection', whereby the rapid expansion of cattle ranching in Brazil also impacted forest cover significantly in the past couple of decades.

Indonesia, another country to display worryingly high deforestation rates, has lost approximately 40 per cent of its forest cover since 1965. The rapid expansion of 'fastwood' plantations of acacia and eucalyptus for pulp and paper, and of palm oil plantations since the early 2000s has put considerable pressure on the archipelago's dwindling forests throughout the past decade.

Another primary cause of deforestation is the increase in transport infrastructure, especially the construction of roads. As Angelsen and Kaimowitz (1999) point out, research clearly shows that greater access to forests and markets accelerates deforestation. Forests in coastal countries and islands are more easily accessible, making them more vulnerable to conversion than in continental countries. Spatial regression models also show a strong correlation between roads and proximity to markets on the one hand, and deforestation on the other, whether in Belize, Honduras, Cameroon or the Philippines. Simpler access to forests makes natural resource exploitation, settlement and land conversion considerably easier. Moreover, the greater the number of access routes, the more difficult it becomes for authorities to monitor the use of the forests.

The role of the timber industry as a driver of deforestation is less clear-cut. Angelsen and Kaimowitz (1999) notably found that the effect of higher timber prices was controversial; on the one hand, higher prices could increase the net benefits of land and encourage deforestation, but on the other, lower prices could discourage efficient harvesting, leading in turn to more logging. Moreover, selective logging, as it is practised in many tropical forests, is conducive to opening access to forests, but is not necessarily a direct cause of deforestation per se.

As for underlying causes of deforestation, none proved to be conclusive. Among them, population pressure is often highlighted as a major factor, but effects cancel themselves out once additional independent variables are added. Likewise, studies suggesting that higher income levels and economic growth have an impact on deforestation produced unclear results, as did those that correlated deforestation with external debt, trade, structural adjustment and technological change. Only in examples of structural adjustment that included pro-agricultural policies have higher deforestation rates been recorded.

Two points can be made from these observations. Firstly, the causes of deforestation are always multiple and complex; it is difficult, if not impossible, to untangle the different causes. Only in a small number of cases can causal relationships be clearly established, notably with agricultural prices and road accessibility.

Secondly – and more importantly – the main causes of deforestation lie outside the forest sector. The timber industry has often taken much of the blame for deforestation in recent decades, especially as images of large trees being felled produce impressive imagery for popular imagination. Unsustainable logging practices have also been widely documented and increasing access to information and communication, including in remote forest areas, have shed light on the effect of logging activities when they are carried out in the absence of silvicultural regulations.

It is time to think outside the box. Researchers have known for decades that other sectors – notably agriculture and transport – can have a much greater impact on forests than the forest sector itself. Yet decision-makers have been slow in realizing the importance of cross-sectoral linkages when addressing the deforestation crisis.

Thinking outside the Box

Today, two major impediments still hamper efforts to address the current forest degradation and deforestation crisis. The first of these is the inability of many to see forests as having multiple values, functions and purposes. These have long been recognized by researchers, but many stakeholders continue to see forests as little more than a source of timber.

In Indonesia, for instance, legislation stipulates that forests containing less than $5m^3$ of commercially viable timber per hectare can be cleared for other land uses. This has opened the door for large-scale conversion of over-logged forests for the benefits of ever-expanding acacia, eucalyptus and palm oil plantations. Likewise, landowners in the Brazilian Amazon often perceive forests as little more than a source of timber, and thus give them less monetary value than what they believe is a more viable alternative, such as cattle ranching. Yet studies have repeatedly shown that if all of forests' values were to be recognized and internalized, conversion to other land uses would not be economically justifiable.

'Internalizing these externalities', however, (such as provision of clean water, prevention of soil erosion, and cultural and spiritual values) is not sufficient; efforts also need to be made at an institutional level. This leads us to the second major impediment: the failure to look beyond the forest sector. This brief overview of the causes of deforestation clearly shows that the forest sector does not stand on its own, but is intimately linked to a range of other sectors impacting land use, notably agriculture and transport.

Both impediments are connected to each other, as stakeholders of a particular sector often see forests from a single perspective. The timber industry has long perceived forests as timber stocks while conservationists saw them as a source of biodiversity. Only recently have some stakeholders, notably in the forest sector, begun acknowledging that forests are home to more than just one value. The gradually increasing uptake of certification labels such as Forest Stewardship Council (FSC) is a clear illustration of this, particularly in the timber industry, considering that such labels are based on the preservation and/or sustainable use of all forest values.

So far, decision-makers and institutions have lagged behind in recognizing both the need for cross-sectoral and cross-institutional linkages and the multiple values of forests. It is time to realize that, unless these issues are addressed in national and international policies, deforestation is likely to continue unabated.

Two promising initiatives are currently being elaborated. The first is institutional cooperation at the international level, notably within the UN system. Since its inception in 2000, the UNFF has promoted a 360-degree perspective for all things related to forests, recognizing the need to widen the debate on deforestation well beyond the forest sector. This was internationally agreed in 2007, when all 192 member states of the UN agreed to address forests in this comprehensive fashion and to implement the four global objectives on forests. This is the first time that an international institution has embraced a holistic approach to combating deforestation and forest degradation. In 2009, the UNFF Secretariat signed a Memoranda of Understanding with a number of UN organizations, notably with the Secretariat of the Convention on Biological Diversity (CBD), thus acknowledging the intricate link between forests and biodiversity.

More recently still, UNFF reached an agreement to work closely with the Secretariat of the United Nations Convention to Combat Desertification (UNCCD) in order to address the institutional gap between drylands and forests, but also in recognition of the protective functions of forests in preventing soil erosion and maintaining clean water supplies. This agreement also draws a direct link between sustainable forest management and sustainable land management, which integrates all land uses, including agriculture, transport, environmental conservation and forestry.

In the past few years, a second approach focused on finance has been emerging, which is equally promising. REDD+ first appeared as an attempt to recognize the value of forests as carbon stocks, but in the past few years it has made its way to the forefront of the international debate as one of the most important instruments for tackling deforestation, forest degradation and forest restoration in developing countries. The great strength of this concept is that it is results-oriented and thus pledges remuneration on reducing deforestation *whatever its causes*. In this sense, it constitutes a genuinely cross-sectoral mechanism as it treats the causes of deforestation equally, whether inside or outside the forest sector.

At the same time, negotiations have also taken place within the UNFF, culminating in the adoption of the Resolution on the Means of Implementation of Sustainable Forest Management in October 2009. This resolution saw the creation of the Facilitative Process and an Ad Hoc Expert Group on forest financing, both of which aim to assist countries in mobilizing new and existing financial resources for forests. The Facilitative Process was immediately kick-started with the launching of a project to identify gaps, obstacles and opportunities to forest financing in Small Island Developing States and low forest cover countries, the two categories to have suffered the most from a decline in forest financing over the past two decades. This project aims to be of cutting-edge quality as it focuses on a cross-sectoral perspective to understand the negative and positive impacts of all sectors on forests.

This 360-degree approach, both institutional and financial in nature, is still in the making, and national and international actors must not stop here. Maintaining forests and reducing deforestation and forest degradation will require an unprecedented level of cooperation that needs to reach well beyond the forest sector. Such an approach requires a strong scientific foundation that upholds the close linkages between sectors and the multiple values of forests. It requires thinking and acting outside the box. It requires unprecedented cooperation between ministries in all countries, many of which continue to function in competition with each other and certainly do not work to meet mutual objectives. We all share a view of the importance of forests, of all the benefits that forests provide, but we do not match our institutional and sectoral actions to our shared vision. Only by doing so will we ensure that forests remain sustainably managed, for present and future generations.

References

Angelsen, A. and Kaimowitz, D. (1999) 'Rethinking the causes of deforestation: Lessons from economic models', *The World Bank Research Observer*, vol 14, no 1, pp73–98

FAO (2010), *Global Forest Resource Assessment 2010*, available at www.fao.org/docrep/013/i1757e/i1757e.pdf

Part III

Climate Change Biodiversity and Ecosystems

8

Climate Change Means a Less Habitable Planet

Thomas E. Lovejoy

Arrhenius's original scientific work that identified the greenhouse effect was actually meant to address the question of why this planet has the particular temperature it does. More specifically, why is it a temperature so favourable to humans and other forms of life instead of being too cold? The difference is the consequence of heat-trapping gases such as carbon dioxide but also methane and others.

What was not known at that time was any detail about the previous global temperature. But that has been revealed by tiny bubbles of air trapped in ice formations that reflect both the temperature and the composition of the atmosphere at the time of entrapment. Ice cores open that record book to science.

We now know that the average temperature of the planet has been unusually stable for the last 10,000 years. Not only does this mean that the entire human enterprise developed during and depends on a stable climate, but also that all the planet's ecosystems have spent the same ten millennia adapting to a stable climate.

The pre-industrial concentration level of carbon dioxide was 280ppm. The current concentration is about 390ppm and will climb to 450ppm if global emissions peak in 2016 (which seems hardly even attainable at this point).

The global climate system has begun to change and the average global temperature is already 0.75°C warmer as a consequence. A fair amount of further change is built into the system.

Probably the most obvious changes we are seeing are those that involve the frozen part of the planet (the cryosphere) and, in particular, the solid and liquid phases of water. Most glaciers are in retreat and all tropical glaciers (on the very high peaks) are retreating at a rate where they will be gone in 15 years. Very dramatic changes are taking place in the northern polar regions; the sea ice expanding and contracting annually on the Arctic Ocean has been retreating to an increasing degree and reached its lowest extent this year. It has also been declining in thickness at the same time.

This has obvious implications for all species with natural histories that depend on ice, the best known of which is the polar bear. Another example would be nesting seabirds. For example, the black guillemot nests on land and the adults fly to the edge of the sea ice to feed on Arctic cod, which never venture far from the edge of the ice. As the edge of the ice has retreated, it eventually becomes too far to fly, leading to nesting colony failure.

This is an example of a decoupling event, in which two aspects of nature are closely timed and interdependent, but one depends on a day length as a cue for timing and the other on temperature. Another example would be the snowshoe rabbit now found in brilliant white winter pelage but no longer camouflaged because, with a warmer spring, the landscape is no longer snowy.

Generally speaking, species are adjusting their annual natural cycles. Many wildflowers in the UK are flowering earlier in the spring. Lilacs are blooming earlier in New England. Animals also are changing life cycle timing with birds migrating, nesting and laying eggs earlier in both Europe and North America.

In addition, species distributions are changing. Edith's checkerspot, one of the best studied butterfly species in North America, occurs in western North America (primarily coastal states). It has clearly been moving upward in altitude and northward in latitude, tracking its required climatic and ecological conditions. A large number of European butterfly species are moving similarly.

Similar distribution shifts are being recorded in aquatic environments. Ocean plankton and fish species are changing in distribution, with obvious implications for fisheries. In the great estuary of eastern North America, the Chesapeake Bay, seagrass has a very strict upper temperature limit, so the southern border of the entire seagrass communities has been moving steadily northward year after year.

The above kinds of change – once exceptional and anecdotal – are now apparent in most places. It is statistically robust: nature is on the move almost everywhere anyone looks in the world.

More disturbing than these relatively minor ripples in nature, problems are emerging such as has recently been documented for tropical lizards (*Sceloparus* and other genera) in different parts of the world. Increasing temperatures have reduced the amount of foraging time so dramatically that female lizards can no longer successfully breed.

In addition, there are a growing number of examples of ecosystem failure. The first of these – and indeed a very dramatic example – involves the fundamental partnership at the heart of tropical coral reef ecosystems, namely between a coral animal and an alga. Only a slight elevation in temperature (even for a short time) is enough for the coral to expel the alga. That causes what are known as 'bleaching events': basically the entire ecosystem fails, and the highly diverse and productive Technicolor world – so important for adjacent human communities – collapses and goes black and white. The first bleaching event occurred in 1983, and they have been occurring with increasing frequency as the world warms.

Another example of ecosystem failure involves the coniferous forests of North America, where milder winters and longer summers have tipped the balance in favour of the native bark beetle. From southern Alaska through a great swathe of the western Canadian provinces and US states, there are places where 70 per cent of the trees are dead as a consequence. Beyond being a major challenge for forest and fire management, it is not at all clear what the future of these ecosystems may be. It is likely distinctly different ecosystems will take their place with real but hard to predict consequences for the original biodiversity.

What is clear in broad terms is that, as global temperature increases, there will be more example such as the lizards, coral reefs or coniferous forests, most of which will be obvious in retrospect and not easy to anticipate. Some of these will come about as the result of unfortunate synergies between climate change and other kinds of stresses being experienced by biodiversity and ecosystems.

An interesting and disturbing example is the prospect of 'Amazon dieback' in which the moisture level necessary to maintain tropical rainforests would fall to a level that they would be replaced by tropical savannah (with great consequences in biodiversity loss, loss of carbon to the atmosphere and serious diminishment of human well-being in the region). This was first recognized as a possibility in about 2005, when the Hadley Centre predicted dieback in the southern and east Amazon if the planet warmed by 2.5°C. In 2009, it redid the model and lowered the tipping point to 2.0°C.

In 2009, the World Bank supported a major study that modelled the combined impact of deforestation, fire and climate change – just as the forest experiences them all together. This indicated the tipping point could occur at as low as 20 per cent deforestation – a breath away from the 18 per cent current deforestation.

This, of course, can be addressed by proactive and aggressive reforestation in the 'arc of destruction' to restore the margin of safety. Disturbingly the third Global Biodiversity Outlook identified 13 tipping points such as tropical coral reefs and Amazon dieback.

There is also system change at a vast scale in changing ocean chemistry. Overlooked by almost everyone until well into the past decade was that some of the immense amounts of carbon dioxide being absorbed by the oceans (and keeping the atmospheric greenhouse gas (GHG) burden much lower than it would otherwise be) was actually being converted into carbonic acid and making the oceans more acidic. Today the oceans are 0.1 pH unit more acid than in pre-industrial times (which is equivalent to 30 per cent more acid since pH has a logarithmic scale).

This also has immense implications for ocean biodiversity. A very large number of ocean species build shells or skeletons from calcium carbonate, derived from a carbonate equilibrium. That in turn is dependent on temperature and pH, so the colder and the more acid the water, the harder it is for calcium carbonate to be mobilized. This has implications for tens of thousands of species. Some of these species are known mostly to science, yet exist in untold numbers at the base of food chains. The impacts are potentially huge, even if dimly perceived and understood at this point.

There are additional complications that lie ahead. Climate change of course is not new in the history of life on Earth and huge glaciers came and went in the northern hemisphere without apparent loss of biodiversity. Species were basically able to track and follow their required conditions. The difference today is that landscapes are largely highly modified by human activities rendering them obstacle courses for species attempting to move in response to climate change. The policy response to the obstacle course challenge is very straightforward: put connections back into the natural landscape – often something that is desirable for other reasons as well.

As ecosystems experience more climate change, it is known from climate change in the geologic past that ecosystems and biological communities do not move as a unit, but rather that it is the individual species that move each at their own rate and in their own direction. The consequence – as has been studied in Europe at the end of the last glaciation – is that the species move and the ecosystems disassemble, with the surviving species assembling into novel associations that are quite hard to predict or manage. In other words, the greater the amount of climate change, the more difficult it is for biodiversity, and the harder to devise practical management approaches.

The conclusion that all this drives toward is that the less the climate changes the better it is for biodiversity and for all the things biodiversity provides for humanity. Indeed it is clear that a rise of 2°C is too much for ecosystems and that fact, together with the consequent sea-level rise, argues strongly for staying below that number.

One number that has been suggested as a desirable target for the concentration of carbon dioxide is 350ppm (and around 1.5°C). The problem, of course, is that the current concentration is around 390ppm and if we are not to exceed 450ppm we need to peak in global emissions in 2016. So that leads inevitably to the question of how to pull some of the carbon dioxide out of the atmosphere.

There is one clear way to do it, namely through the power of biology. Indeed, twice in the history of life on Earth there have been very high levels of carbon dioxide brought down to levels akin to pre-industrial. The first was the time of the origin of plants on land, and the second was the time of the spread of modern flowering plants. In both instances, the drawdown was not caused just by the green plants and photosynthesis, but also by the biodiversity in the soil, which assisted the green plants and their photosynthesis through provision of nutrients. That same soil biodiversity augmented soil formation which itself sequesters atmospheric carbon dioxide in the process.

The difference between then and now is we cannot afford to wait millions of years for the process. It turns out, however, that human activities (deforestation, habitat conversion of various sorts etc.) have been responsible for releasing perhaps 200 billion or more tons of carbon to the atmosphere over the past three centuries. And a significant amount of that could be recaptured.

A programme of ecosystem restoration at the planetary scale has the potential to pull about 1.5 billion tons of carbon as carbon dioxide out of the atmosphere or about 40ppm over a 50-year period – the difference between the 390ppm of today and the target 350ppm. This would be annually about 0.5 billion tons in reforestation, 0.5 billion tons in restoring degraded grasslands and grazing land, and another 0.5 billion tons from managing agriculture in ways that restores carbon to the soils. To give a sense of scale; the 0.5 billion tons a year that could be sequestered by forests could be done with reforestation in the tropics of an area equivalent to the size of Spain (for the most optimistic scenario) or an area three times larger (for the less optimistic). This would simultaneously make all those ecosystems more resilient to climate change and other stresses they will encounter.

This does not, of course, reduce the imperative of rapidly and radically transforming the energy base of human society into carbon neutrality. Even that cannot happen fast enough to avoid dangerous human-caused change, so it argues for a crash research programme to identify ways to physically remove additional carbon dioxide from the atmosphere and convert it into some inert substance. That would be an acceptable form of 'geo-engineering' without the high risk that most geo-engineering inherently contains. Most geo-engineering, in any case, addresses temperature only (a symptom) rather than carbon dioxide concentrations (the cause).

In the end, all must recognize the planet works as a biophysical system and that anything we do with the physical elements, such as raising the temperature, has of necessity to be viewed through the biological lens. In the end, we depend on the biological infrastructure of the planet and it has enormous, virtually endless, potential to benefit humanity – as long as it remains robust. The planet has to be managed as a biophysical system, and the place to start is to 'regreen' the Emerald Planet and use the living planet to make the planet more habitable.

9

REDD+ as a Potentially Viable Solution to Biodiversity and Ecosystem Insecurity

Yemi Katerere, Linda Rosengren, Lera Miles and Matea Osti

Introduction

Tropical forests cover more than 600 million hectares of the Earth's surface and contain at least two thirds of the world's terrestrial biodiversity (Gardner et al, 2009). Forests not only benefit humankind with a vast number of products – both timber and non-timber – but they also provide ecosystem services such as regional and local climate regulation, water supply and soil conservation. Tropical forests also serve as significant carbon sinks, a vital ecosystem service, storing more than half of all carbon found in terrestrial vegetation worldwide (Watson et al, 2000).

In recent decades, forests and the biodiversity that they nurture have come under increasing pressure from activities associated with land-use change, primarily deforestation and forest degradation. The prognosis is not good; although the global rate of deforestation is showing signs of decreasing, it is still alarmingly high, with around 13 million hectares of forest lost each year over the last decade (FAO, 2010). A 2010 report by the Secretariat of the Convention on Biological Diversity (Secretariat of the CBD, 2010) concluded that the biodiversity target agreed to in 2002 by governments to 'achieve by 2010 a significant reduction of biodiversity loss' had not been met and that the principal pressures driving biodiversity loss, including habitat change or loss from deforestation, were either staying constant or increasing. These pressures have helped turn forests into considerable sources of anthropogenic greenhouse gas (GHG) emissions, with 6–17 per cent of all emissions attributed to land-use change resulting mainly from tropical forest loss and degradation (van der Werf et al, 2009)

The focus on forests at the international climate negotiation level has increased substantially in recent years, as recognition of the role that forests play in influencing the global carbon balance and their potential as a climate change mitigation measure has grown. Reducing Emissions from Deforestation and Forest Degradation and the role of conservation, sustainable management of forests and enhancement of forest carbon stocks (REDD+) has been proposed as a mechanism under the United Nations Framework Convention on Climate Change (UNFCCC) to contribute to the mitigation of global climate change by maintaining and enhancing the forests in developing countries, many of which contain high levels of biodiversity and provide an abundance of ecosystem services. Five activities have been proposed under REDD+: reducing emissions from deforestation, reducing emissions from forest degradation, conservation of forest carbon stocks, sustainable management of forests and enhancement of forest carbon stocks.

In addition to its primary goal of climate change mitigation by avoiding emissions from the forestry sector, the REDD+ mechanism, if successfully implemented, could provide other significant benefits. This chapter discusses these additional benefits from REDD+, also known as multiple benefits. We first outline the concept of multiple benefits and examples of measures to safeguard and enhance them. Next, we discuss one potentially effective and appropriate method of safeguarding and enhancing multiple benefits under a REDD+ mechanism, known as payments for ecosystem services (PES). We then analyse the role of monitoring as a mechanism for improving our understanding of the impact of REDD+ activities on multiple benefits. Finally, we present the key challenges facing REDD+, with suggestions as to how they might be overcome.

REDD+ and Multiple Benefits

Multiple benefits

Maintaining natural carbon stocks through mechanisms such as REDD+ can generate multiple benefits, in other words, additional benefits to the primary aim of climate change mitigation. Examples include biodiversity conservation and the ecosystem services that are direct outcomes from protecting and maintaining natural ecosystems. There are other types of multiple benefits that derive from the mechanisms used and the social and political changes needed to implement them, such as the clarification of land tenure and enhanced stakeholder participation in decision-making. The types, combinations and scale of multiple benefits vary according to the approach used and geographical locations being considered.

Multiple benefits in international negotiations

The multiple benefits of REDD+ (and some of the potential risks) have been receiving growing attention in the ongoing REDD+ negotiations in the UNFCCC. The latest draft negotiating text under the UNFCCC lists the safeguards that should be promoted and supported, including actions that help to incentivize the protection and conservation of natural forests and their ecosystem services, and actions that enhance other social and environmental benefits (UNFCCC, 2010).

Other policy processes have also identified the ecosystem benefits derived from REDD+. The parties to the Convention on Biological Diversity (CBD) passed a decision that noted that REDD+ could provide multiple benefits for biodiversity and reducing GHG emissions (CBD, 2008). This same decision also established an Ad Hoc Technical Expert Group on Biodiversity and Climate. The report of the Ad Hoc Technical Expert Group noted that REDD+ would have a positive impact on biodiversity conservation (Secretariat of the CBD, 2009).

Measures to safeguard and enhance multiple benefits

The outcomes of multiple benefits in any particular case will be shaped by the ecological characteristics of the forests, the types of REDD+ activities in question and how those activities are implemented. The outcomes will also be influenced by any measures that are adopted to enhance benefits or safeguard against ecosystem harm.

If a successful REDD+ mechanism is implemented, tropical forest biodiversity is expected to benefit, mainly through incentives to reduce the loss of forest (Miles and Kapos, 2008; Venter et al, 2009; Harvey et al, 2010), which is currently the single major threat to biodiversity in the tropics (Ravindranath, 2007). One benefit to biodiversity and ecosystem services can include the avoidance of losses if more biodiversity is retained with a REDD+ project or programme than without one. Benefits may also include improvements or enhancements to the present situation. For example, if REDD+ projects are implemented with landscape connectivity in mind, this is likely to enhance benefits to biodiversity and ecosystem services by maximizing species mobility and habitat range and optimizing certain ecological processes, such as water regulation.

Payments for ecosystem services

One type of mechanism that might be useful for safeguarding and enhancing multiple benefits is an incentive-based approach known as payments for ecosystem services (PES). PES schemes transfer money from the recipients of an ecosystem service to those responsible for its maintenance or enhancement. In some countries, REDD+ itself may be implemented through a PES-type scheme, with performance-based payments compensating forest owners and users for forest carbon conservation. Where willing funders are available, or legislation compels payments by service users, additional payments may be made for non-carbon services of forests to produce 'premium' REDD+ credits. This would have positive effects on those ecosystem services that are prioritized by a country and valued under such a scheme. To date, PES schemes tend to fund water provision, soil conservation and biodiversity conservation. Wunder (2007) provides a useful review of PES and clarifies its scope as a tool for tropical conservation. In a separate review, Bond et al (2009) concluded that PES can contribute to REDD+, provided that certain institutional, economic, informational and cultural conditions are met up front.

A new approach to contract landowners to supply ecosystem services is being tested in Australia (Australian Government, 2006). Landowners are given the opportunity to receive funding to maintain and restore the biodiversity of valued ecosystems. They do so by bidding for a conservation contract or 'agreement' in a 'reverse auction', which identifies a set of sites from the available bids that offer the greatest conservation value for the least cost. The conservation value is evaluated using a 'conservation value index' that includes the significance of the conservation feature (its size and quality), the long-term security of the commitment to protect it, and the service provided by the landowner (the management of the conservation asset). When the auction round is complete, the remainder of those landowners who have made a bid are given a 'take it or leave it' offer, based on the average value of the agreements for successful bids with an equivalent conservation value. Landowners commit to specific conservation actions and must refrain from converting the native vegetation for a number of years or in perpetuity because a restrictive covenant to the landholding is attached. Multiple auction rounds have been offered to date, an iterative process that has enabled learning by both the government and landowners and has also established a market price for the conservation agreements.

The model integrates individual landholders, a nationally managed fund and multiple criteria to assess the quality of bids – such as multiple benefit provision – into a market mechanism. Prerequisites include clarity of tenure, a regulatory system that allows long-term covenants to be placed on land, a monitoring system with redress for non-compliance and excellent outreach to landholders. It would be worth assessing whether this type of model could be adapted for use in REDD+ eligible countries. An analogous conservation value index could also be formulated for REDD+, but rigorous testing would be required to assess suitability for the desired outcomes.

Monitoring multiple benefits

The primary reason for monitoring multiple benefits is to understand the impact of REDD+ activities upon them and to adapt the activities if necessary to safeguard and enhance the benefits. This process does not necessarily require any reporting to or verification by a third party. However, a monitoring process for one or more benefits will most likely be required if a PES scheme is in place, or if premium REDD+ credits are being sold.

In terms of cost, it would make sense to link the monitoring of ecosystem benefits to the carbon monitoring system if the data and methodology permit. While costs may limit the scope for additional monitoring of ecosystem benefits, it is crucial that REDD+ plans identify the most important ecosystem benefits to monitor, rather than the easiest.

Monitoring is only useful when there is a clear purpose – it should not be undertaken for its own sake. It has been suggested that ecosystem benefits need to be monitored in order to obtain regular feedback on the implementation and impacts of readiness activities (Global Witness, 2010). It has also been argued that if ecosystem benefits are not measured, there is no way to ensure that REDD+ activities, such as forest restoration, are leading to the delivery of ecosystem services (Palmer and Filoso, 2009).

Challenges in Making REDD+ Work

There is no question that REDD+ is a potentially powerful policy instrument for influencing how tropical forests are managed (Stickler et al, 2009). Furthermore, REDD+ could provide opportunities for synergies between environmental and social benefits (Okereke and Dooley, 2010). However, several challenges with regard to the design and implementation of the REDD+ regime remain to be solved.

The *raison d'être* of REDD+ is to achieve carbon emission reductions from forests. However, forests harbour an estimated 75 per cent of all terrestrial biodiversity and provide a wide range of ecosystem goods and services that are essential for human well-being. Nonetheless, some argue that the formal REDD+ mechanism should not be overburdened by requiring these additional benefits and that the focus should remain on the main objective of REDD+: carbon emission reduction. It has been noted, however, that the capacity of forests to resist external pressures, including climate change, is dependent on their biodiversity (Thompson et al, 2009). This suggests that biodiversity serves a more fundamental role as an enabling condition for the long-term success of REDD+. There are also concerns that a singular focus on carbon will drive policy objectives to focus on emission reductions without considering the broader causes of deforestation (Okereke and Dooley, 2010).

REDD+ activities may harm biodiversity and ecosystem services, and it is also possible that REDD+ activities may have positive effects on some ecosystem services and negative effects on others. The outcomes depend on how the activities are implemented and under what circumstances. The REDD+ activity that has given rise to the greatest concerns is the enhancement of forest carbon stocks, particularly in the form of afforestation or reforestation. The development of plantation forests could lead to the loss of biodiversity that was formerly present and to the reduction in water regulatory services. A successful REDD+ mechanism may also have indirect negative impacts through indirect land-use change, such as affecting food security. By reducing the conversion rate of forests and by increasing afforestation and reforestation, land that might otherwise have been available for agriculture could no longer be available. This, in turn, may increase the conversion of other ecosystem types – such as savannah and wetlands – to agriculture, with the consequent loss of the biodiversity and ecosystem services provided by those lands (Miles and Kapos, 2008). It is possible to address at least some of these potential risks through the appropriate design and implementation of REDD+. The safeguards currently under negotiation in the UNFCCC offer one way to ensure that this happens.

Leakage presents an additional challenge to the implementation of REDD+. The problem of leakage occurs when deforestation moves from one area to another within a country or between countries without a net emission reduction (Ghazoul et al, 2010). National and regional REDD strategies, together with proper enforcement and coordination should reduce the incidence of leakage.

Another major concern is the issue of 'permanence' – the stability over time of the REDD+ achievements (Ghazoul et al, 2010). It will be crucial to continue to provide incentives to curb deforestation after the first REDD+ payments have been received in order to accomplish long-term emission reduction.

Expectations that REDD+ can benefit all forests and countries equally need to be carefully managed. REDD+ mechanism is essentially broadened to not only address deforestation and forest degradation but to include the conservation of forest carbon stocks, the sustainable management of forests and the enhancement of forest carbon stocks. Also of concern will be the reference scenario, which is historical, or projected emissions. While the broadening of the scope of REDD+ could increase the carbon emission reduction potential of the mechanism and offer more countries the possibility to participate in and benefit from the REDD+ regime, carbon benefits are unlikely to be equitably distributed. This is because asymmetries in carbon density exist among countries. For instance, Brazil and Indonesia together account for 45 per cent of emissions from deforestation (Okereke and Dooley, 2010) and are likely to be the major REDD+ beneficiaries. Broadening the scope of the REDD+ mechanism could offer countries with historically low deforestation rates the possibility of benefiting from REDD+ and hence create greater geographical equity in the REDD+ regime.

Asymmetries in carbon density also exist within a country. Therefore, equitable distribution of the financial benefits that the REDD+ mechanism will provide is also a fundamental question to be addressed at a country level. Payments should, as the mechanism is designed now, go to forest landowners. This requires that countries address tenure rights and in many cases customary rights in their national REDD+ strategies (Ghazoul et al, 2010).

The opportunity to link environmental and social issues under REDD+ can benefit from the lessons and experiences learned from past integrated conservation and development projects (ICDPs), which sought to combine the goals of conservation and development at the project level (Blom et al, 2010). Since many national REDD+ programmes are likely to start with sub-national projects, ICDPs can offer valuable lessons for REDD+ on what has or has not worked at the project level.

Conclusions

REDD+ presents an opportunity to address many of the challenges related to the underlying causes of global tropical forest cover loss. If planned carefully,

REDD+ could symbiotically contribute to biodiversity conservation, the provisioning of ecosystem services and the enhancement of human well-being, in concert with its primary goal of climate change mitigation.

However, as has been demonstrated, the extent of multiple benefits from REDD+ depends on a variety of factors, from the geographical and institutional components of a given country to the architectural elements of the REDD+ mechanism. PES provides one example of a model that could be integrated into a REDD+ framework that allows the multiple benefits of REDD+ to be realized. As REDD+ implementation moves forward, determining the effectiveness of REDD+ in maintaining and enhancing multiple benefits will rely on comprehensive monitoring to understand the actual benefits accrued over time, which will also allow for subsequent REDD+ planning to be readjusted according to the specific needs of future sites and regions.

As the REDD+ regime evolves, awareness of the complexity of the mechanism will help improve its design and implementation. A key consideration will be minimizing those aspects of the REDD+ mechanism that may act to undermine or exclude multiple benefits and supporting those activities that enhance them.

References

Australian Government (2006) *Strategic Plan for the Forest Conservation Fund*, www.environment.gov.au/land/publications/forestpolicy/pubs/strategic-plan-fcf.pdf

Blom, B., Sunderland, T. and Murdiyarso, D. (2010) 'Getting REDD to work locally: Lessons learned from integrated conservation and development projects', *Environmental Science & Policy*, vol 13, no 2, pp164–172

Bond, I., Grieg-Gran, M., Wertz-Kanounnikoff, S., Hazlewood, P., Wunder, S. and Angelsen, A. (2009) 'Incentives to sustain forest ecosystem services: A review and lessons for REDD', Natural Resource Issues no 16, International Institute for Environment and Development, London, with Center for International Forestry Research, Bogor, Indonesia, and World Resources Institute, Washington, DC

CBD (Convention on Biological Diversity) (2008) *Decision IX/16 Biodiversity and Climate Change*, Convention on Biological Diversity, www.cbd.int/decisions/cop/?m=cop-09

FAO (Food and Agriculture Organization) (2010) *Global Forest Resources Assessment 2010: Key Findings*, Food and Agriculture Organization, Rome

Gardner, T. A., Barlow, J., Chazdon, R., Ewers, R. M., Harvey, C. A., Peres, C. A. and Sodhi, N. S. (2009) 'Prospects for tropical forest biodiversity in a human-modified world', *Ecology Letters*, vol 12, no 6, pp561–582

Ghazoul, J., Butler, R. A., Mateo-Vega, J. and Pin Koh, L. (2010) 'REDD: A reckoning of environment and development implications', *Trends in Ecology & Evolution*, vol 25, no 7, pp396–402

Global Witness (2010) *Review of JPDs and R-PPs submitted to the 4th UN-REDD Policy Board and 5th FCPF Participants Committee Meetings: Provisions on Enforcement and Non-carbon Monitoring*, Global Witness, London

Harvey, C. A., Dickson, B. and Kormos, C. (2010) 'Opportunities for achieving biodiversity conservation through REDD', *Conservation Letters*, vol 3, no 1, pp53–61

Miles, L. and Kapos, V. (2008) 'Reducing greenhouse gas emissions from deforestation and forest degradation: Global land-use implications', *Science*, vol 320, no 5882, pp1454–1455

Okereke, C. and Dooley, K. (2010) 'Principles of justice in proposals and policy approaches to avoided deforestation: Towards a post-Kyoto climate agreement', *Global Environmental Change*, vol 20, no 1, pp82–95

Palmer, M. A. and Filoso, S. (2009) 'Restoration of ecosystem services for environmental markets', *Science*, vol 325, no 5940, pp575–576

Ravindranath, N. H. (2007) 'Mitigation and adaptation synergy in forest sector', *Mitigation and Adaptation Strategies for Global Change*, vol 12, no 5, pp843–853

Secretariat of the CBD (Convention on Biological Diversity) (2009) *Connecting Biodiversity and Climate Change Mitigation and Adaptation: Report of the Second Ad Hoc Technical Expert Group on Biodiversity and Climate Change*, CBD Technical Series no 41, Secretariat of the Convention on Biological Diversity, Montreal

Secretariat of the CBD (2010) *Global Biodiversity Outlook 3*, Secretariat of the Convention on Biological Diversity, Montreal

Stickler, C. M., Nepstad, D. C., Coe, M. T., McGrath, D. G., Rodrigues, H. O., Walker, W. S., Soares-Filho, B. S. and Davidson, E. A. (2009) 'The potential ecological costs and cobenefits of REDD: A critical review and case study from the Amazon region', *Global Change Biology*, vol 15, no 12, pp2803–2824

Thompson, I., Mackey, B., McNulty, S. and Mosseler, A. (2009) 'Forest resilience, biodiversity, and climate change: A synthesis of the biodiversity/resilience/stability relationship in forest ecosystems', CBD Technical Series no 43, Secretariat of the Convention on Biological Diversity, Montreal

UNFCCC (United Nations Framework Convention on Climate Change) (2010) *Text to Facilitate Negotiations Among Parties*, FCCC/AWGLCA/2010/6, http://unfccc.int/resource/docs/2010/awglca10/eng/06.pdf

van der Werf, G. R., Morton, D. C., DeFries, R. S., Olivier, J. G. J., Kasibhatla, P. S., Jackson, R. B., Collatz, G. J. and Randerson, J. T. (2009) 'CO_2 emissions from forest loss', *Nature Geoscience*, vol 2, no 11, pp737–738

Venter, O., Meijaard, E., Possingham, H., Dennis, R., Sheil, D., Wich, S., Hovani, L. and Wilson, K. (2009) 'Carbon payments as a safeguard for threatened tropical mammals', *Conservation Letters*, vol 2, no 3, pp123–129

Watson, R. T., Noble, I. R., Bolin, B., Ravindranath, N. H., Verardo, D. J. and Dokken, D. J. (eds) (2000) *Land Use, Land-use Change, and Forestry: A Special Report of the Intergovernmental Panel on Climate Change*, Cambridge University Press, Cambridge

Wunder, S. (2007) 'The efficiency of payments for environmental services in tropical conservation', *Conservation Biology*, vol 21, no 1, pp48–58

10

Water and Biodiversity

Georgia Destouni and Johan Kuylenstierna

Introduction

Water is directly or indirectly linked to most development and environmental challenges, making its role important to address in connection with biodiversity and ecosystem status and change. The Fourth Assessment Report of the Intergovernmental Panel on Climate Change (IPCC) has sent a clear message on the role of water in relation to climate change: 'If mitigation is about energy, adaptation is about land and water' (IPCC, 2007). The projected impacts of climate change will hit ecosystems first and foremost through the water cycle and through changes at the landscape scale. Land and water changes are therefore central components for understanding and interpreting climate change effects on biodiversity and ecosystems.

However, while climate and land-use changes are commonly in focus when trying to understand and project fundamental drivers behind ecosystem change and loss of biodiversity, changes in the water variable – for example, due to overexploitation, pollution or infrastructure development – are not often discussed. Nevertheless, both the direct water impacts (due to availability limitations, floods and droughts) and its indirect role as a link and propagator of change between different drivers (climate change, water uses of different economic sectors and water pollution), and their downstream ecosystem effects are important if we intend to seriously address biodiversity and ecosystem decline.

Increasing competition over water is one of the most severe threats to ecosystem security and biodiversity, as it is for sustainable development. For instance, the Secretariat of the Convention on Biological Diversity (CBD) has provided a comprehensive overview of the role of water, wetlands and forests and their ecological, economic and policy linkages (Blumenfeld et al, 2009). Therefore, when approaching the water linkage to ecosystems and biodiversity development, it is essential to do so through a wider development lens. In this chapter, we have elected to focus in particular on the food security and climate change aspects of global development and water; mainly because food production and food security in a changing climate represents one of the most significant challenges faced by humanity and they are so closely related to both water and biodiversity development.

Food Security, Water and Climate Change

The Food and Agricultural Organization (FAO) states that food security 'is the outcome of food system performance at global, national and local levels' (FAO, 2008a). Its management requires a systems approach, as it is 'directly or indirectly dependent on agricultural and forest ecosystem services, e.g. soil and water conservation, watershed management, combating land degradation, protection of coastal areas and mangroves, and biodiversity conservation'. This definition clearly incorporates the relationships between climate, water, ecosystems and biodiversity change.

In order to understand changes in the pressures on water resources, and associated impacts on ecosystems and biodiversity, it is clearly essential to consider food production and food security trends simply because agriculture is the dominant consumer of water. Over the past century, global food production has managed to match population growth. Despite a threefold global population increase since the turn of the 1900s, global production is still enough to sustain 6.5 billion people, even if such indicators as the ratio of global cereal stocks to utilization has been declining or remained stable at a rather low level since the late 1990s (for most recent data, refer to www.fao.org).

The increases in agricultural output in the 20th century can be attributed to horizontal expansion of arable land, the capacity to intensify production through the application of seed, fertilizer and pesticide technologies, and the ability to pump, store and divert surface and groundwater. Such factors were largely behind the 'green revolution', a period characterized by significant increases in agricultural output in most parts of the world, and notably in countries such as India and China. Dams, water diversions and other infrastructure harnessing water resources (from lakes, rivers and groundwater) facilitated farming and energy production. In addition, increasing trade enabled food to be transported from surplus countries and regions to countries and regions that did not have enough food production capacity and/or chose to allocate land and water resources to other productive uses. As a result, it is estimated that 60 per cent of the large river systems in the world have been moderately or strongly affected (MEA, 2005) and water withdrawals have tripled since 1950.

FAO projects that a combination of future population growth and economic growth will push food requirements to double the current levels by 2050 (FAO, 2006), including an increase of grain production from 2 billion tons to more than 4 billion tons. Current food production consumes more than 2500 billion cubic metres of water annually, or 75 per cent of total freshwater consumption (FAO Aquastat Database, 2008). This level of demand will have far-reaching consequences for the allocation and redistribution of water resources from various ecosystem and biodiversity needs to different productive economic sectors.

Numerous recent publications point to the anticipated impacts of climate change on water and agriculture (IPCC, 2007; World Bank, 2007; Bates et al, 2008; FAO, 2008a). Climate change impacts on rain-fed agriculture are transmitted through soil moisture deficits and temperature increases, while impacts on irrigated agriculture are transmitted through the overall availability of water resources. Irrigation is in itself a driver of regional climate change (Boucher et al, 2004; Shibuo et al, 2007; Lobell et al, 2009). Furthermore, even if a rain-fed and an irrigated agricultural system are subjected to the same set of human demand drivers (population growth, income growth), the factors of supply and the points of competition over water resources are quite different. Rain-fed agriculture does not compete for rainfall, while irrigated agriculture competes for water with other productive sectors and natural ecosystems.

The links between climate, water and food production are complex, but the relation between temperature, water and plant physiology is essentially fixed. For any C3 (such as wheat) or C4 plant (such as maize), a fixed amount of evapotranspiration and carbon dioxide is required to assimilate carbon (Steduto et al, 2007). Put simply, more food or fibre production requires more soil water – whether it is derived from rainfall or from surface and groundwater sources through irrigation. While 'more crop per drop' may be an objective for overall irrigation management and delivery of water to the soil, it is clear that any increase in biomass can only be attained through increased water availability in the soil. While climate determines what can be grown at any particular location, it is the range of hydrological changes anticipated under various climate change scenarios that gives the main causes for concern.

Regions already struggling with complex food-related challenges (marginal areas, subsistence farming, poverty, competition, management challenges, and so on) will clearly be sensitive to climate and hydrological change. The larger agricultural systems, such as the areas of continuous irrigation in Asia, may be more buffered in terms of water availability and ability to apply technology. However, basin-wide changes in temperature, evapotranspiration and water availability will here have greater impacts on the global food supply. Assessing the impacts on various scales of the interlinked changes in climate, hydrology, global food production and biodiversity is, therefore, a key scientific challenge.

Although uncertain, IPCC has made assessments of possible climate change effects. In Africa, 75–250 million people are projected to be exposed to increased water stress, and yields from agriculture are expected to decrease as much as 50 per cent in some countries while the area of semi-arid and arid land increases. Land areas classified as very dry have already doubled since the 1970s (Bates et al, 2008). In Asia, freshwater availability in many large rivers may decrease and changes in water availability from glacier and snow melting will affect water availability and thus also agriculture and biodiversity. In the Middle East, scenario projections of high significant temperature increases suggest an expected 3.5 per cent loss in gross domestic product (GDP) due to loss of arable land and threats to coastal cities (FAO, 2008c). In Latin America, there may be gradual replacement of tropical forests by savannah, and productivity of some important crops is projected to decrease. Lobell et al (2008) point out that South Asia and Southern Africa are two regions with food production based on crops that are likely to be negatively affected by climate change. However, the effects are in the end also strongly dependent upon changes in socio-economic parameters, with the projected range of increasing numbers of hungry people in the future being very wide (Schmidhuber and Tubiello, 2007).

Climate change impacts are not only confined to developing countries. Agriculture and forestry is expected to become increasingly difficult in eastern Australia as aridity intensifies. In Europe, the already significant regional differences in water availability are expected to increase, with drought becoming even more common in the Mediterranean region. North America is expected to increase its rain-fed agriculture in the eastern and northern parts, while decreasing snow and ice will reduce summer flows in already water scarce western regions. Kerr (2008) presents potential hot spots in North America, providing an interesting overview of associated challenges. In general, decisions related to agriculture and food security in the changing climate will have important implications for water, ecosystem and biodiversity changes on a worldwide scale.

Responding to Challenges

If a growing population is to be fed and the volatility of rain-fed systems adequately buffered to maintain global food security, only the delivery of more water into the root zone of more productive land and more efficient use of available food resources can assure the required increased production and availability. Increased food production and associated increases in land and water use will affect the status of ecosystems and biodiversity. In so far as the latter constitute measures of the health of planet Earth, then the patient is getting worse. Butchart et al (2010) concluded based on 31 biodiversity related indicators that most trends are negative. This is serious, not least in light of the commitments made in 2002 by 188 governments within the Convention on Biological Diversity (CBD) to considerably reduce the rate of biodiversity loss by 2010. As a consequence, managing and controlling interlinked climate, water, agricultural and biodiversity changes will have to continue to be priority issues for policy and decision-makers in the coming decades.

Over the recent decades, various global crises have been proclaimed one after the other. Although there may be a tendency to exaggerate the impact of each individual emergency, it is becoming increasingly clear that global challenges are progressively becoming more complex and interlinked. Perhaps it is now time to acknowledge that, rather than just dealing with occasional crises, we need to understand and manage the dynamics of continuous environmental change. There is already a tendency towards greater environmental management flexibility and adaptability, rather than trying to find blueprint solutions. A management culture that promotes adaptive capacity to deal with continuous change, rather than only with occasional crises, is necessary.

Policies and actions related to changes in climate, water, agriculture and biodiversity clearly need to be better incorporated into existing key development processes. Goals for environmental management have so far commonly been defined relative to conserving or restoring some perceived 'natural' conditions. However, there are major difficulties in defining the natural state, modelling how a system might be progressed toward such a natural state and not least with the feasibility and actual desirability of restoring a natural state (Bishop et al, 2009). Failure to critically examine the complexities of having 'natural' as the goal for environmental and biodiversity management will compromise the ability to manage the full complexity of global development issues, which is unfortunate both for the Western world embracing a single model of 'natural as the goal' and for the developing world if they are uncritically encouraged to just adopt this model. A necessary complementary environmental management model must involve a process of continuously navigating though ever-changing environmental and socio-economic conditions, steered by long-term development objectives.

In this navigation process, open access to relevant national and international information is a fundamental prerequisite for improved policy-making and management, enabling us to cope with and adapt to the ongoing changes (Hannerz et al, 2005; Arctic-HYDRA, 2010). Furthermore, openly accessible monitoring and scientific data, and state-of-the-art knowledge need to be translated into policy and management relevant information that can be of direct significance to decision-making at various levels.

Competition over scarce resources must also be further addressed. Ecosystems too often end up as the last one in line when resources are allocated. Increasing competition between food production and energy is likely to be a particularly critical case. Increases in biofuel production have a direct impact on water consumption and food availability. Although biofuels could be a potential for many poor countries, areas already experiencing water stress, or those on the brink of it, could see reduced water availability for the more basic needs of people as well as for vital ecosystems. As Varghese (2007) states: 'The indiscriminate promotion of biofuel development as a "cheap and green" energy option may interfere with optimal water allocation, and/or the pursuit of appropriate public water policies that will help address the water crisis.' Although biofuel feedstock currently accounts for only 1 per cent of the total area under tillage, and a similar percent of crop water use, production is likely to grow rapidly. Impacts are still poorly understood. Demand for biofuels based on agricultural feed stocks will be a significant factor over the coming decades and it has already contributed to higher food prices (FAO, 2008b).

In relation to climate change impacts on water, agriculture and biodiversity, provision of relevant information requires an increased focus on change projections, knowledge transfer and capacity-building at the user level. For a farmer, urban planner or water resources manager, projected global climate change averages over large spatial and temporal scales are not of any real practical use. The capacity to make projections at regional and local scales needs to be strengthened, and information disbursement is required to strengthen the capacity of users to interpret and use such information. As stated by the FAO (2007), 'improved access to knowledge is only theoretical for many in poor countries especially in rural areas' as long as efficient technologies, including the internet, are not available. Development of a range of methods to share knowledge at different user levels is therefore needed.

There is also a need to integrate climate change-related challenges with those of other types of regional and global change (Howden et al, 2007). Land-use and water-use changes, large-scale water diversions, economic development, changes in consumption and production patterns (agriculture, industry), changes in population and population dynamics will all influence water availability and quality, food security and biodiversity. In many cases, socio-economic and other environmental changes may eclipse the local–regional manifestations of short to medium-term climate change (Darracq et al, 2005; Destouni and Darracq, 2009). A carefully measured application of science and economics is needed to review and understand different linkages and feedback systems, and a better understanding of the linkages and feedbacks must form the foundation for relevant and effective policy interventions.

Acknowledgement

Parts of this chapter were based on an unpublished background document prepared for the sixth World Water Forum held in Turkey in March 2009, by Jakob Burke (FAO) and Johan Kuylenstierna (UN-Water). The views expressed in this publication are those of the authors and do not necessarily reflect the views of the FAO, UN-Water or the Stockholm University. The designations employed and the presentation of material in this information product do not imply the expression of any opinion whatsoever on the part of the FAO concerning the legal or development status of any country, territory, city or area or of its authorities, or concerning the delimitation of its frontiers or boundaries.

References

Arctic-HYDRA (2010) *The Arctic Hydrological Cycle Monitoring, Modelling and Assessment Programme: Science and Implementation*, Grafia Kommunikasjon AS, Oslo

Bates, B. C., Kundzewicz, Z. W., Wu, S. and Palutikof, J. P. (eds) (2008) 'Climate change and water', IPCC Technical Paper no VI, Secretariat of the Intergovernmental Panel on Climate Change, Geneva

Bishop K., Beven K., Destouni G., Abrahamsson K., Andersson L., Johnson R., Rodhe J. and Hjerdt N. (2009) 'Nature as the "natural" goal for water management: A conversation', *AMBIO: A Journal of the Human Environment*, vol 38, no 4, pp209–214

Blumenfeld, S., Lu, C., Christophersen, T. and Coates, D. (2009), 'Water, wetlands and forests: A review of ecological, economic and policy linkages', CBD Technical Series no 47, Secretariat of the Convention on Biological Diversity and Secretariat of the Ramsar Convention on Wetlands, Montreal and Gland

Boucher, O., Myhre, G. and Myhre, A. (2004) 'Direct human influence of irrigation on atmospheric water vapour and climate', *Climate Dynamics*, vol 22, nos 6–7, pp597–603

Butchart, S. H. M., Walpole, M., Collen, B., van Strien, A., Scharlemann, J. P. W., Almond, R. E. A., Baillie, J. E. M., Bomhard, B., Brown, C., Bruno, J., Carpenter, K. E., Carr, G. M., Chanson, J., Chenery, A. M., Csirke, J., Davidson, N. C., Dentener, F., Foster, M., Galli, A., Galloway, J. N., Genovesi, P., Gregory, R. D., Hockings, M., Kapos, V., Lamarque, J. F., Leverington, F., Loh, J., McGeoch, M. A., McRea, L., Minasyan, A., Hernández Morcillo, M., Oldfield, T. E. E., Pauly, D., Quader, S., Revenga, C., Sauer, J. R., Skolnik, B., Spear, D., Stanwell-Smith, D., Stuart, S. N., Symes, A., Tierney, M., Tyrrell, T. D., Vié, J-C. and Watson, R. (2010) 'Global biodiverity: Indicators of recent declines', *Science*, vol 328, no 5982, pp1164–1168

Darracq, A., Greffe, F., Hannerz, F., Destouni, G. and Cvetkovic, V. (2005) 'Nutrient transport scenarios in a changing Stockholm and Mälaren valley region, Sweden', *Water Science and Technology*, vol 51, nos 3–4, pp31–38

Destouni G., and Darracq, A. (2009) 'Nutrient cycling and N_2O emissions in a changing climate: The subsurface water system role', *Environmental Research Letters*, vol 4, no 3, 035008

FAO (Food and Agriculture Organization) (2006) *State of Food Insecurity in the World*, Food and Agriculture Organization, Rome

FAO (2007) *Adapting to Change on Our Hungry Planet: FAO at Work 2006–2007*, Food and Agriculture Organization, Rome

FAO (2008a) *Climate Change and Food Security: A Framework Document*, Food and Agriculture Organization, Rome

FAO (2008b) *Climate Change, Water and Food Security: The State of Food and Agriculture, Biofuels: Prospects, Risks and Opportunities*,
Food and Agriculture Organization, Rome

FAO (2008c) 'Twenty-ninth FAO regional conference for the Near East', Conference Document: NERC/08/INF/5, Cairo, 1–5 March 2008

FAO Aquastat Database (2008) *Aquastat Database*, www.fao.org/nr/water/aquastat/main/index.stm

Hannerz, F., Destouni, G., Cvetkovic, V., Frostell, B. and Hultman, B. (2005) 'A flowchart for sustainable integrated water management following the EU Water Framework Directive', *European Water Management Online*, 2005/04, www.ewaonline.de/journal/2005_04.pdf

Howden, S. M., Soussana, J-F., Tubiello, F. N., Chhetri, N., Dunlop, M. and Meinke, H. (2007) 'Adapting agriculture to climate change', *PNAS*, vol 104, no 50, pp19691–19696

IPCC (Intergovernmental Panel of Climate Change) (2007) 'Climate change: Impacts, adaptation and vulnerability', *Contribution of WGII to the Fourth Assessment Report*, Cambridge University Press, Cambridge

Kerr, R. (2008) 'Climate change hot spots mapped across the United States', *Science*, vol 321, no 5891, p909

Lobell, D. B., Burke, M. B., Tebaldi, C., Mastrandrea, M. D., Falcon, W. P. and Naylor, R. L. (2008) 'Prioritizing climate change adaptation needs for food security in 2030', *Science*, vol 319, no 5863, pp607–610

Lobell, D., Bala, G., Mirin, A., Phillips, T., Maxwell, R. and Rotman, D. (2009) 'Regional differences in the influence of irrigation on climate', *Journal of Climate*, vol 22, no 8, pp2248–2255

MEA (Millennium Ecosystem Assessment) (2005) *Ecosystems and Human Well-being: Synthesis*, Island Press, Washington, DC

Schmidhuber, J. and Tubiello, F. N. (2007) 'Global food security under climate change', *PNAS*, vol 104, no 50, pp19703–19708

Shibuo, Y., Jarsjö, J. and Destouni, G. (2007) 'Hydrological responses to climate change and irrigation in the Aral Sea drainage basin', *Geophysical Research Letters*, vol 34, L21406

Steduto, P., Hsiao, T. C. and Fereres, E. (2007) 'On the conservative behavior of biomass water productivity', *Irrigation Science*, vol 25, no 3, pp189–207

Varghese, S. (2007) *Biofuels and Global Water Challenges*, Institute for Agriculture and Trade Policy, Minnesota

World Bank (2007) 'Agriculture for development', *World Development Report 2008*, World Bank, Washington, DC

11

Mitigating Climate Change through Ecological Agriculture

Lim Li Ching

Introduction

The Intergovernmental Panel on Climate Change (IPCC) warns that warming of the climate system is 'unequivocal', as is evident by increases in air and ocean temperatures, widespread melting of snow and ice, and sea-level rise (IPCC, 2007a). Agriculture will therefore have to cope with increased climate variability and more extreme weather events. A recent report warns that unchecked climate change will have major negative effects on agricultural productivity, with yield declines for the most important crops and price increases for the world's staples – rice, wheat, maize and soybeans (Nelson et al, 2009).

The IPCC projects that crop productivity would increase slightly at mid to high-latitudes for local mean temperature increases of up to 1–3°C (depending on the crop) (Easterling et al, 2007). However, at lower latitudes, especially in the seasonally dry and tropical regions, crop productivity is projected to decrease for even small local temperature increases (1–2°C). In some African countries, yields from rain-fed agriculture, which is important for the poorest farmers, could be reduced by up to 50 per cent by 2020 (IPCC, 2007b). Further warming above 3°C would have increasingly negative impacts in all regions.

The number of people at risk of hunger will therefore increase, although impacts may be mitigated by socio-economic development. Overall, however, the assessment is that climate change will affect food security in all its dimensions – food availability, access to food, stability of food supplies and food utilization (FAO, 2009).

The impacts of climate change will fall disproportionately on developing countries, despite the fact that they contributed least to the causes. Furthermore, the majority of the world's rural poor who live in areas that are resource-poor, highly heterogeneous and risk-prone will be hardest hit by climate change. Smallholder and subsistence farmers, pastoralists and artisanal fisherfolk will suffer complex, localized impacts of climate change and will be disproportionately affected by extreme climate events (Easterling et al, 2007). For these vulnerable groups, even minor changes in climate can have disastrous impacts on their livelihoods (Altieri and Koohafkan, 2008).

Agriculture's Contribution to Climate Change

While agriculture will be adversely affected by climate change, it also contributes to the problem. Agriculture directly releases into the atmosphere a significant amount of carbon dioxide, methane and nitrous oxide, amounting to around 10–12 per cent or 5.1–6.1 gigatonnes of carbon dioxide equivalent per year (Gt CO_2-eq/yr) of global anthropogenic greenhouse gas (GHG) emissions annually (Smith et al, 2007). More current estimates put the figure at 14 per cent or 6.8 Gt CO_2-eq/yr (FAO, 2009).

Of global anthropogenic emissions in 2005, agriculture accounted for about 58 per cent of nitrous oxide and about 47 per cent of methane (Smith et al, 2007), both of which have far greater global warming impact than carbon dioxide. Nitrous oxide emissions from agriculture are mainly associated with nitrogen fertilizers and manure applications, as fertilizers are often applied in excess and not fully utilized by crops, such that some surplus is lost to the atmosphere. Fermentative digestion by ruminant livestock contributes to agricultural methane emissions, as does cultivation of rice in flooded conditions.

However, if indirect contributions (such as land conversion to agriculture, synthetic fertilizer production and distribution and farm operations) are factored in, it is estimated that the contribution of agriculture could be as high as 17–32 per cent of global anthropogenic emissions (Bellarby et al, 2008). In particular, land-use change, driven by industrial agricultural production methods, would account for more than half of total (direct and indirect) agricultural emissions. Deforestation to expand arable land and poor agricultural soil management lead to significant carbon dioxide emissions as carbon stocks above and below ground are depleted (IFOAM, 2009).

Conventional industrial agriculture is also heavily reliant on fossil fuels. The manufacture and distribution of synthetic fertilizers contributes a significant amount of GHG emissions, between 0.6–1.2 per cent of the world's total (Bellarby et al, 2008). This is because the production of fertilizers is energy intensive and emits carbon dioxide, while nitrate production also generates nitrous oxide.

Future emissions growth

Total GHG emissions from agriculture are expected to increase, reaching 8.3 Gt CO_2-eq/yr in 2030 (Smith et al, 2007). If food demands increase and dietary shifts occur as projected, then annual agricultural emissions may rise further.

Agricultural emissions of nitrous oxide are projected to increase 35–60 per cent up to 2030 due to increased nitrogen fertilizer use and increased animal manure production, while methane emissions related to global livestock production are also projected to increase by 60 per cent up to 2030 (FAO, 2003). Direct emissions of carbon dioxide from agriculture are likely to decrease or remain low; however, indirect causes, such as converting land to agriculture, would contribute substantial emissions.

At a regional level, the highest growth in emissions from agriculture are projected for sub-Saharan Africa, together with the Middle East and north Africa, with a combined 95 per cent increase in the period 1990–2020 (Smith et al, 2007). This is linked to rising demand for livestock products and intensification and expansion of agriculture to still largely unexploited areas, particularly in south and central Africa.

Agriculture's Mitigation Potential: Making the Case for Ecological Agriculture

Although agriculture is a significant contributor to climate change, it also has considerable mitigation potential. The IPCC estimates that the global technical mitigation potential from agriculture by 2030 to be about 5.5–6.0 $GtCO_2$-eq/yr (Smith et al, 2007), with soil carbon sequestration being the mechanism responsible for most (89 per cent) of the mitigation potential. Therefore, agriculture could potentially change from being one of the largest GHG emitters to a much smaller emitter and even a net carbon sink (Bellarby et al, 2008).

There are a variety of practices that can reduce agriculture's contribution to climate change. These include crop rotations and improved farming system design, improved cropland management, improved nutrient and manure management, improved grazing-land and livestock management, maintaining fertile soils and restoration of degraded land, improved water and rice management, fertilizer management, land-use change and agroforestry (Smith et al, 2007; Bellarby et al, 2008; Niggli et al, 2009).

These practices essentially entail a shift to more sustainable farming that builds up carbon in the soil and uses less chemical fertilizers and pesticides (ITC and FiBL, 2007; Bellarby et al, 2008). Many of these techniques are already common practice in what can be termed as 'ecological agriculture'. Ecological agricultural approaches, including organic agriculture, generally integrate natural, regenerative processes, minimize non-renewable inputs (pesticides and fertilizers), rely on the knowledge and skills of farmers and depend on locally adapted practices to innovate in the face of uncertainty (Pretty and Hine, 2001).

Ecological agriculture fosters biodiversity and is in itself biodiverse – not only in terms of the harvested elements (both intra- and interspecies genetic diversity) but also in terms of the components necessary to maintain the agroecosystem (Ensor, 2009) (see Box 11.1). It depends on and sustains agricultural biodiversity and has come about through the innovation of farmers over time. Biodiverse agriculture mimics nature and works with nature, in contrast to conventional industrial agriculture, which tends to simplify agricultural systems and reduce diversity.

Box 11.1 Common characteristics of biodiverse farms

- Species and structural diversity are combined in time and space through vertical and horizontal organization of crops.
- Higher biodiversity of plants, microbes and animals supports crop production and mediates a reasonable degree of biological recycling of nutrients.
- The full range of micro-environments is exploited.
- Effective recycling practices maintain cycles of materials and waste.
- Biological interdependencies provide some level of biological pest suppression.
- Reliance on local resources plus human and animal energy.
- Reliance on local varieties of crops and incorporation of wild plants and animals.

Source: Adapted from Ensor (2009), based on Altieri and Koohafkan (2008)

Ecological agriculture, and organic agricultural systems in particular, have inherent potential to reduce emissions and to enhance carbon sequestration in soils (Scialabba and Müller-Lindenlauf, 2010). The total mitigation potential of organic agriculture has been estimated at 4.5–6.5 $GtCO_2$-eq/yr, with potentially much higher amounts possible depending on agricultural management practices (Muller and Davis, 2009). The financial requirements are low, as carbon sequestration and low-emissions farming can be achieved through inexpensive means (IFOAM, 2009), which are immediately available and can be implemented without long delays in research and development (Scherr and Sthapit, 2009).

Reducing emissions

Agricultural soils can be managed to reduce emissions by minimizing tillage, reducing the use of nitrogen fertilizers and preventing erosion (Scherr and Sthapit, 2009). Practising organic agriculture also reduces nitrous oxide and methane emissions from biomass waste burning (which accounts for about 12 per cent of agricultural emissions), as burning is avoided (Muller and Davis, 2009). Moreover, organic standards ban the certification of recently cleared or altered primary ecosystems such as forests, slowing emissions from forest conversion to agriculture (IFOAM, 2009), while the use of catch and cover crops in organic systems prevents soil erosion and hence soil carbon loss.

In particular, the careful management of nutrients and hence the reduction of nitrous oxide emissions from soils – the most important source of agricultural emissions – is a significant contribution of organic agriculture (Scialabba and Müller-Lindenlauf, 2010). Approximately 20 per cent of agricultural emissions could be reduced by converting to organic agriculture, through its omission of synthetic nitrogen fertilizers – 10 per cent due to lower energy demand as a result of avoiding emissions incurred during fertilizer production, and 10 per cent due to lower nitrous oxide emissions as a result of lower nitrogen input than in conventional agriculture (Niggli et al, 2009; Scialabba and Müller-Lindenlauf, 2010).

Nitrogen input in ecological agriculture instead comes from the application of manure and compost, or is provided by the focus on agricultural biodiversity, in particular rotations that include legumes (ITC and FiBL, 2007; Ensor, 2009). In addition, catch and cover crops extract plant-available nitrogen that was unused by the preceding crop, reducing the amount of reactive nitrogen in the topsoil and hence nitrous oxide emissions (Ensor, 2009; Scialabba and Müller-Lindenlauf, 2010).

Soil carbon sequestration

The highest mitigation potential of ecological agriculture lies in carbon sequestration in soils. The technical potential of carbon sequestration in world soils may be 2–3 billion tonnes per year for the next 50 years (Lal, 2009). Carbon sequestration is encouraged by practices that leave residues and reduce tillage to encourage build up of soil carbon. Increasing the role of perennial crops and agroforestry further allows carbon storage while crops are being produced (Ensor, 2009; Lal, 2009; Scherr and Sthapit, 2009). While these strategies are not exclusive to ecological agriculture, they clearly resonate with ecological principles.

For example, ecological agriculture practices such as crop rotation, cover crops, manuring and application of organic amendments such as compost restore degraded soils and hence increase soil carbon sequestration (Scialabba and Müller-Lindenlauf, 2010). Ecological agriculture also stresses the importance of maintaining and enhancing biodiversity (for example, field margins, hedges, trees or bushes), which are effective mitigation strategies, due to carbon sequestration in soil and plant biomass.

It is estimated that a conversion to organic agriculture would considerably enhance the sequestration of carbon in soils. Organic systems have been found to sequester more carbon than conventional farms (ITC and FiBL, 2007; Bellarby et al, 2008; Niggli et al, 2009). Niggli et al (2009) estimate that a conversion to organic farming would mitigate 40 per cent (2.4 Gt CO_2-eq/yr) of the world's agriculture GHG emissions, in a minimum scenario, or up to 65 per cent (4 Gt CO_2-eq/yr) in a maximum scenario (including no-tillage) of carbon sequestration. Other estimates point to higher potentials of 6.5–11.7 Gt CO_2-eq/yr (Muller and Davis, 2009).

Nonetheless, the increase in soil organic matter eventually reaches equilibrium and the mitigation effect can be reversed if the carbon stored is released (for example, by ploughing of no-tillage systems) (Scialabba and Müller-Lindenlauf, 2010). While the total sequestration capacity of soils is finite, there are an estimated 50–100 years of remaining sequestration potential (Smith et al, 2007).

Concurrent Benefits for Adaptation

Ecological agriculture optimally integrates mitigation and adaptation, as many of its approaches that mitigate climate change are also effective adaptation strategies.

Soil carbon sequestration is a clear example of a mitigation measure that also enhances adaptation and the sustainability of crop production (Smith, 2009). The increased soil organic matter enhances soil fertility and quality, improves water-holding capacity and increases productivity and resilience, which are important for adaptation to future climate change (Lal, 2009). In particular, ecological agriculture practices such as crop rotations, composting, green manures and cover crops can reduce the negative effects of drought while increasing productivity (ITC and FiBL, 2007; Niggli et al, 2009). Organic matter also enhances water capture in soils, significantly reducing the risk of floods (ITC and FiBL, 2007; Niggli et al, 2009).`

Agricultural biodiversity is the keystone of ecological agriculture. It contributes to mitigation – diverse plants and trees in crop rotations and in the surrounding agroecosystem sequester carbon, while the incorporation of legumes reduces nitrous oxide emissions. On the other hand, resiliency to climate disasters is closely linked to agricultural biodiversity. Practices that enhance biodiversity allow farms to mimic natural ecological processes, enabling them to better respond to change and reduce risk. Thus, farmers who increase interspecific diversity suffer less damage during adverse weather events, compared to conventional farmers planting monocultures (Altieri and Koohafkan, 2008; Ensor, 2009; Niggli et al, 2009). Moreover, the use of intraspecific diversity (different cultivars of the same crop) is insurance against future environmental change. Diverse agroecosystems can also adapt to new pests or increased pest numbers (Ensor, 2009).

Other examples of coincident mitigation and adaptation strategies include: application of animal manure, which reduces fertilizer use, improves soil structure and water-holding capacity; reduction of tillage intensity with improved residue management, which increases soil carbon while retaining soil moisture; and restoring degraded lands, which sequester carbon and enhance soil resilience (Smith, 2009).

Conclusion

Climate change will undoubtedly pose serious challenges for agriculture. However, with appropriate focus on ecological agriculture to provide mitigation, adaptation and increased productivity options, a 'win-win-win' scenario for agriculture is possible. This is because ecological agriculture would not only be beneficial in terms of climate mitigation and adaptation, but would also constitute the paradigm shift in agriculture that is deemed necessary to increase productivity while ensuring sustainability and meeting smallholder farmers' food security needs (IAASTD, 2009).

Therefore there is a clear need to invest more resources, research and training into ecological agriculture, as well as to provide the appropriate policy and funding support (IAASTD, 2009). Many components of ecological agriculture can also be applied to improve all farming systems, including conventional ones. A crucial factor would be investment in the conservation, protection and enhancement of agricultural biodiversity, which underpins ecological approaches in agriculture.

Maximizing the synergies between mitigation and adaptation in ecological agriculture means that these strategies should be developed simultaneously:

- Further research is needed on the mitigation and adaptation options provided by ecological agriculture, taking into account context and location specificities such as soil types, crop types, management practices and climate conditions.
- Arrangements should be made for the sharing of information and experiences, transfer of and training in good practices that constitute mitigation and adaptation in ecological agriculture, including through extension services.
- Countries should urgently adopt and implement mitigation and adaptation action plans for agriculture, focusing in particular on ecological agriculture.
- Financing assistance for adaptation and mitigation measures in the agriculture sector in developing countries should be prioritized, especially if they constitute ecological agriculture practices.

References

Altieri, M.A. and Koohafkan, P. (2008) 'Enduring farms: Climate change, smallholders and traditional farming communities', TWN Environment and Development Series no 6, Third World Network, Penang

Bellarby, J., Foereid, B., Hastings, A. and Smith, P. (2008) *Cool Farming: Climate Impacts of Agriculture and Mitigation Potential*, Greenpeace International, Amsterdam

Easterling, W. E., Aggarwal, P. K., Batima, P., Brander, K. M., Erda, L., Howden, S. M., Kirilenko, A., Morton, J., Soussana, J. F., Schmidhuber, J. and Tubiello, F. N. (2007) 'Food, fibre and forest products', in M. L. Parry, O. F. Canziani, J. P. Palutikof, P. J. van der Linden and C. E. Hanson (eds) *Climate Change 2007: Impacts, Adaptation and Vulnerability: Contribution of Working Group II to the Fourth Assessment Report of the Intergovernmental Panel on Climate Change*, Cambridge University Press, Cambridge, pp273–313

Ensor, J. (2009) *Biodiverse Agriculture for a Changing Climate*, Practical Action Publishing, Colchester

FAO (Food and Agricultural Organization) (2003) *World Agriculture: Towards 2015/2030: An FAO Perspective*, Food and Agricultural Organization, Rome

FAO (2009) *Climate Change and Bioenergy Challenges for Food and Agriculture*, Food and Agricultural Organization, Rome

IAASTD (International Assessment of Agricultural Knowledge, Science and Technology for Development) (2009) *Agriculture at a Crossroads*, Island Press, Washington, DC

IFOAM (International Federation of Organic Agriculture Movements) (2009) *High Sequestration, Low Emission, Food Secure Farming. Organic Agriculture: A Guide to Climate Change and Food Security*, International Federation of Organic Agriculture Movements, Bonn

IPCC (Intergovernmental Panel on Climate Change) (2007a) 'Climate change 2007: Synthesis report', Contribution of Working Groups I, II and III to the Fourth Assessment Report of the Intergovernmental Panel on Climate Change, IPCC, Geneva

IPCC (2007b) 'Summary for policymakers', in M. L. Parry, O. F. Canziani, J. P. Palutikof, P. J. van der Linden and C. E. Hanson (eds) *Climate Change 2007: Impacts, Adaptation and Vulnerability: Contribution of Working Group II to the Fourth Assessment Report of the Intergovernmental Panel on Climate Change*, Cambridge University Press, Cambridge, pp7–22

ITC and FiBL (International Trade Centre and Research Institute of Organic Agriculture) (2007) *Organic Farming and Climate Change*, International Trade Centre, United Nations Conference on Trade and Development/World Trade Organization, Geneva

Lal, R. (2009) 'The potential for soil carbon sequestration', in G. C. Nelson (ed) *Agriculture and Climate Change: An Agenda for Negotiation in Copenhagen*, Brief 5, Focus 16, International Food Policy Research Institute, Washington, DC

Muller, A. and Davis, J. S. (2009) *Reducing Global Warming: The Potential of Organic Agriculture*, Rodale Institute and Research Institute of Organic Agriculture, US and Switzerland

Nelson, G. C., Rosegrant, M. W., Koo, J., Robertson, R., Sulser, T., Zhu, T., Ringler, C., Msangi, S., Palazzo, A., Batka, M., Magalhaes, M., Valmonte-Santos, R., Ewing, M. and Lee, D. (2009) *Climate Change: Impact on Agriculture and Costs of Adaptation*, International Food Policy Research Institute, Washington, DC

Niggli, U., Fließbach, A., Hepperly, P. and Scialabba, N. (2009) *Low Greenhouse Gas Agriculture: Mitigation and Adaptation Potential of Sustainable Farming Systems*, Food and Agriculture Organization, Rome

Pretty, J. and Hine, R. (2001) *Reducing Food Poverty with Sustainable Agriculture: A Summary of New Evidence*, University of Essex Centre for Environment and Society, Essex

Scherr, S. J. and Sthapit S. (2009) 'Mitigating climate change through food and land use', Worldwatch Report no 179, Worldwatch Institute, Washington, DC

Scialabba, N. E. and Müller-Lindenlauf, M. (2010) 'Organic agriculture and climate change', *Renewable Agriculture and Food Systems*, vol 25, no 2, pp158–169

Smith, P. (2009) 'Synergies among mitigation, adaptation and sustainable development', in G. C. Nelson (ed) *Agriculture and Climate Change: An Agenda for Negotiation in Copenhagen*, Brief 9, Focus 16, International Food Policy Research Institute, Washington, DC

Smith, P., Martino, D., Cai, Z., Gwary, D., Janzen, H., Kumar, P., McCarl, D., Ogle, S., O'Mara, F., Rice, C., Scholes B. and Sirotenko, O. (2007) 'Agriculture', in B. Metz, O. R. Davidson, P. R. Bosch, R. Dave and L. A. Meyer (eds) *Climate Change 2007: Mitigation. Contribution of Working Group III to the Fourth Assessment Report of the Intergovernmental Panel on Climate Change*, Cambridge University Press, Cambridge and New York, p499–532

12

Land Degradation and Exploitation and Its Impact on Biodiversity and Ecosystems

Luc Gnacajda

Land Degradation and the Security Challenge

In the broadest sense, conserving land and water equates to securing our common future. Desertification, land degradation and drought (DLDD) threaten human security by depriving people of their means of life – by taking away food, access to water, the means for economic activities and even their homes. In worst-case scenarios, they undermine national and regional security, force people to leave their homes and can trigger low or high-level intensity conflicts. Threats to soil security unleashed by desertification, land degradation and the effects of drought constitute a peril to securing our common future.

The multidimensional security challenges for biodiversity posed by DLDD operating at different scales may negatively reinforce each other threatening the survival of billions of people, particularly in the drylands of our world. In the past century, the world population tripled and water consumption increased six times. There is a high probability that: climate change impacts; growing water stress; biodiversity loss; and increasing DLDD[1] may all contribute to future food crises, unless extraordinary and innovative strategies, policies and measures are launched now. As such, the land degradation issue can be treated as cumulative – and representing a fundamental biodiversity security challenge. In most cases management and catastrophic natural phenomena drive this degradation. DLDD can be characterized as a human-induced, natural continuum having negative effects on ecosystem functions such as storage and cycling of water and soil resources. DLDD occurs not only in arid, semi-arid and dry sub-humid zones but also has adverse effects on other ecosystems.

Land degradation can be considered in terms of the loss of actual or potential productivity or utility as a result of natural or anthropogenic factors.

At its core, it is the decline in land quality, biodiversity services or reduction in its productivity. In this context of productivity, land degradation results from a mismatch between land quality and land use. Recent work (funded by the Food and Agriculture Organization (FAO)) based on Earth observation data has identified 24 per cent of the global land surface as degrading (Bai et al, 2008). Analysis of the datasets under the Global Land Degradation Assessment (GLADA) study based on 23-year sets of normalized difference vegetation index (NDVI) data reveals a declining trend mainly in Africa south of the Equator, southeast Asia, south China, northern-central Australia, the Pampas, and the boreal forest in Siberia and North America. Almost one fifth of degrading land is cropland, which is more than 20 per cent of all cultivated areas. Some areas of historical land degradation have been so degraded that they are now statistically stable and exist at very low levels of productivity.[2]

Important among physical processes that initiate land degradation are a decline in soil structure leading to crusting, compaction, erosion, desertification, anaerobism, environmental pollution and unsustainable use of natural resources. Significant chemical processes include acidification, leaching, salinization, decrease in cation retention capacity and fertility depletion. Biological processes include reduction in total and biomass carbon, and decline in terrestrial biodiversity. Soil structure is an important property that affects all three degradative processes. Thus, land degradation should be viewed as a biophysical process driven by socio-economic and political causes.

About 33 per cent of the global land surface (42 million square kilometres) is subject to desertification (the deserts or hyper-arid areas not included). The semi-arid to weakly arid areas of Africa are particularly vulnerable, as they have fragile soils, localized high population densities and generally a low-input form of agriculture. The region is affected and, if not addressed, the quality of life of large sections of the population will be affected; in 2002, about 25 per cent of the land was subject to water erosion and about 22 per cent to wind erosion. More than 45 per cent of Africa was affected by desertification, 55 per cent of which is at high or very high risk (GEO3 Factsheet on Africa, 2002). In 'Land resource stresses and desertification in Africa' (Reich et al, 2001) it has been estimated at 46 per cent. 'The significance of this large area becomes evident when one considers that about 43% of the continent is characterized as extreme deserts' (Reich et al, 2001). Many of these countries cannot afford losses in agricultural productivity. Loss of terrestrial biodiversity is partly due to the trend towards specialized production systems and inadequate land use/land planning that would favour the matching of land use with land potential and thereby promote diverse landscapes and products and adapted land-use systems.

Impacts on Increasing Security Needs of the Land

All the elements required for ecosystem services depend on soil, and soil biodiversity is the driving force behind their regulation. Soil and agricultural biodiversity and its future security remain threatened by: (a) the spatial expansion of existing deserts; (b) the severe degradation of soils and related fertility and biodiversity losses due to processes of geophysical, wind and water erosion; and (c) drought resulting in bad harvests and crop yield declines. In degraded and highly insecure landscapes, basic ecosystem services are challenged, especially water as well as food production and supply. Soil security is achieved when efforts succeed to conserve soil fertility by containing land degradation and desertification and when the consequences of drought are reduced by improving people's livelihoods and well-being.

One conclusion for the global environment is that land degradation presents a complex picture of some areas becoming worse, a few getting better, but in aggregate a massive impact on the productivity of the world's soil resources, the natural capital of biodiversity (water availability, water quality, forest biomass, soil fertility, topsoil, inclement microclimates) and on the lives of the rural poor. Of concern are the fundamental linkages between degrading land resources, declining biodiversity, emissions of greenhouse gases (GHGs) from soil and reductions in fixed carbon.

Although planting trees is often advocated to control global warming through carbon dioxide fixation, far more organic carbon is accumulated in the soil. The loss of soil biodiversity reduces the ability of soils to regulate the composition of the atmosphere and counteract global warming. The global soil organic carbon pool is estimated at 1550 gigatonnes (Gt), 73–79Gt of which (around 5 per cent) are stored in Europe (Schils et al, 2008). It is estimated that the total annual cost of erosion from agriculture in the US is about US$44 billion per year, or about US$247 per hectare of cropland and pasture. On a global scale, the annual loss of 75 billion tons of soil costs the world about US$400 billion per year, or approximately US$70 per person per year (Lal, 1998).

Few attempts have been made to assess the global economic impact of erosion. Information on the economic impact of land degradation by different processes on a global scale is not available yet. The economic impact of land degradation on agricultural biodiversity is extremely severe in densely populated South Asia, and sub-Saharan Africa. The productivity of some lands in Africa has declined by 50 per cent as a result of soil erosion and desertification. If accelerated erosion continues unabated, yield reductions by 2020 may be 16.5 per cent. It should be underscored that the economic importance of the contribution of aggregated biodiversity to ecosystem resilience (the capacity of an ecosystem to absorb shocks and stresses in constructive ways) is probably very high but still poorly quantified, although studies have analysed aspects such as the contribution of crop diversity to agricultural yields and farm income (Birol et al, 2005; Di Falco and Perrings, 2005). This important gap in economic assessments of land degradation reflects the difficulty of first quantifying the risks of a system collapse from an ecological perspective, and then measuring people's willingness to pay to reduce those risks that are not yet well understood.

In order to allow for performing cost–benefit analyses for measures to protect soil biodiversity, some economic estimates of the ecosystem services delivered by soil biodiversity need to be provided. At the global level, the improper management of soil biodiversity worldwide has been estimated to cause a loss of US$1 trillion per year (Pimentel et al, 1997).

Consequences for Well-being

Climate change is likely to have significant impacts on all services provided by soil biodiversity. It will typically result in higher carbon dioxide concentrations in the air, modified temperatures and precipitation rates, all of which will modify the availability of soil organic matter. It has been well-substantiated that biodiversity is important for the long-term net sequestration of carbon. However, the composition and biological complexity of plants and soil organisms can have various direct effects on the amount, speed and stability of carbon sequestration. Biodiversity can also affect carbon sequestration indirectly, through the provision of other benefits to society, thereby influencing peoples' willingness to maintain a certain land-use or protection regime. Biodiversity is not just a fortunate by-product of carbon sequestration, but should be viewed as a key intervening factor for land security, without which carbon cycles are unsustainable. Therefore, biodiversity warrants incorporation into the design, implementation and regulatory framework of carbon sequestration initiatives.

Land degradation due to soil, water and wind erosion, fertility and biodiversity loss is prevailing in drylands affecting primarily marginalized people. Due to drought and desertification, each year 12 million hectares are lost where 20 million tons of grain could have been grown (UNEP, 2007). In land-based economies, land degradation and economic growth or lack of it (poverty) are intractably linked; people living in the lower part of the poverty spiral are in a weak position to provide the stewardship necessary to sustain biodiversity. As a consequence, they move further down the poverty spiral, and a vicious cycle is set in motion. Urgent remedial action is essential because species loss and ecosystem degradation are inextricably linked to human well-being. Proper financial value should be attributed to the soil and to the work of those who tend it. A European Commission Joint Research Centre (JRC) report in 2009 determined soil biodiversity to be 'of immense economic importance' and claimed that the monetary value of ecosystem goods and services provided by soils, soil biota and the associated terrestrial systems was estimated in 1997 to be US$13 trillion (Gardi and Jeffrey, 2009).

The protection and enhancement of the role of biodiversity as evolutionary capital can provide a number of option values for local communities in the future. The past focus on protected areas needs to be maintained but with a clearer vision on the linkages between biodiversity and human needs. To this effect, the integration between protected areas and production areas needs to be more consistently advocated. The mutual benefits in so doing are essential to human security and society at large. Examples are numerous: protected mangroves as nurseries for fisheries; forests and scrublands as sources of pollinators; capture and purification of drinking and irrigation water by well-covered watersheds; protected coastal vegetation and mangroves as a buffer against sea intrusions. The key scientific and technical question now is how biodiversity can be integrated into other areas of human endeavor to support ecosystems and livelihoods.

The Millennium Ecosystem Assessment (MEA) (UNEP, 2005) referred to close linkages among desertification, global climate change and biodiversity loss, where desertification contributes to climate change and may become irreversible as a result of climate change.[4] The study noted that understanding the impacts of desertification on human well-being requires that we improve our knowledge of the interactions between socio-economic factors and ecosystem conditions. While the state and extent of land degradation and its impacts on biodiversity loss are important, an understanding of the direct and indirect drivers is also needed, as well as an understanding of how society responds and livelihoods are impacted. Climate change is already creating a more difficult and problematic production scenario, especially for small-scale farmers in the world's drylands, where the increasing prevalence of drought and floods cause wide-scale food insecurity.

There are many, usually confounding, reasons why land users permit their land to degrade and biodiversity to deteriorate. Degradation is also a slow imperceptible process and many people are simply not aware that their land is degrading. Creating awareness and building up a sense of stewardship are important steps in the challenge of reducing degradation. A striking aspect of the consequences of biodiversity loss is its disproportionate impact on the poor. Many of the reasons are related to societal perceptions of land and the values they place on land. Many researchers have noted that subsistence farmers, the rural poor and traditional societies face the most serious risks from land degradation (Srinivasan et al, 2008). Consequently, appropriate technology and payment for ecosystem services are only partial answers. The main solution lies in the behaviour of the farmer who is subject to the economic and social pressures of the community and country in which they live. The consequences of biodiversity loss and ecosystem service degradation – from water to food to fish – are not being shared equitably across the world. The areas of richest biodiversity and ecosystem services are in developing countries, where they are relied upon by billions of people to meet their basic needs. This imbalance is likely to grow.

Soil biodiversity, as a primordial ecosystem supporting service, is virtually a *non-renewable* resource. It is a basic tenet of sustainable development that society has an obligation to protect soil, conserve it or even enhance its quality for future generations. The role of society in sustaining agricultural biodiversity can be demonstrated, and, conversely, the role of soil in sustaining society. The paradigms that brought some countries of the world to agricultural affluence must be evaluated, as well as the policies and practices that have contributed to land degradation and decline in productivity in other countries. We need to look at current concerns and the urgent need to develop new paradigms for managing soil resources that will carry us through the next few decades.

Restoring Biodiversity through Sustainable Land Management

Since the 2005 MEA, there has been a better understanding of the biophysical and socio-economic trends relating to land degradation in global drylands. Science is playing the lead role in highlighting new threats to the global environment from the loss of biodiversity due to land degradation. The severity of the problem and the options for slowing down the acceleration of land degradation, the mitigation of climate change, the management of water resources within and across national boundaries and the prevention of chemical pollution must all be based on the best available science and technology.

The protection of the biodiversity of dry and sub-humid lands is important in combating land degradation and desertification. It can also provide income-generating opportunities for dryland communities and contribute to poverty eradication, although it is often necessary to support such opportunities through activities such as improving access to markets, providing payments for ecosystem services and establishing labelling for sustainably harvested products. Furthermore, the conservation of locally adapted species of plants and animals can increase the resilience of the ecosystem in the face of drought. For example, droughts have been demonstrated to have a more significant impact on imported livestock species when compared to local varieties or wild relatives (although this may not affect their relative productivity over the long term). The development and transmission of traditional, scientific and technological knowledge of the interconnections between land degradation and biodiversity through policy-relevant research is crucial. This should be combined with a rapid translation of usable content into the education curriculum and training of experts. Science and knowledge-based policy strategies are an utmost priority for coping with biodiversity loss and boosting resilience in the face of DLDD.

While natural hazards (drought) cannot be prevented, processes of land degradation and desertification can be mitigated by proactive human activities. The impact of DLDD on biodiversity can be reduced by linking protection with empowerment of the people to become more adaptive and resilient. People can be a major asset in reversing a trend towards degradation and further biodiversity loss. However, they need to be healthy and politically and economically motivated to care for the land, as subsistence agriculture, poverty and illiteracy can be important causes of land and environmental degradation. As land resources are essentially non-renewable, it is necessary to adopt a generalized rehabilitation approach to sustainable management of finite resources. The complex cyclical interactions and feedbacks between the Earth and human systems that determine climate change, water stress and biodiversity loss have different effects on DLDD. The often chaotic interrelations can have unpredictable consequences on societal outcomes.

Making Choices

Land is a limited natural resource and land degradation will remain an important global issue for the 21st century because of its adverse impact on agronomic productivity and biodiversity, as well as its effect on food security and the quality of life. Good governance, scientific recognition and public awareness call for processes of anticipatory learning, risk management and proactive policies to mitigate the probable societal impacts of the complex nature–human interactions in order to prevent the projected trends from becoming a future reality. DLDD and its downward spiral effect on biodiversity loss poses multiple threats to international, national and human security that may overstretch both the classic security policy of the state and the capacities of the global environmental governance system.

The proposed recommendations for post-2010 biodiversity targets are based on some optimism that rates of extinction are controllable, habitats can be successfully managed and that biodiversity will be successfully integrated into national policies. However, a renewed focus on land degradation and soil security as a key biodiversity strategy will surely enhance problem awareness on environmental risks for the population and ecosystems.[5] The priority should be on how best to maintain the sustained flow of global environmental benefits through conservation, restoration and incorporation of a new environmental security paradigm in the design of production systems (GEF-STAP, 2010).

For biodiversity loss in the face of land degradation, renewed efforts to measure and address the impacts of degradation processes on the functioning of ecosystems are essential, requiring the development of better monitoring methods and better understandings of the drivers of land degradation and deforestation. A high priority should be given to mainstreaming biodiversity in interventions related to climate change adaptation and mitigation, land degradation and international waters. In many cases, this will involve biodiversity not simply as a trade-off or positive side effect, but as an environmental security tool and means to more effectively achieve the primary goal, the conservation of life on earth in its widest sense.

The purpose of the United Nations Convention to Combat Desertification (UNCCD) through its ten-year strategy (2008–2018) is to reverse and prevent desertification/land degradation and to mitigate the effects of drought in affected areas in order to support poverty reduction and environmental sustainability, in other words, to guide us towards more enlightened land (and water) use and how best to protect and improve the livelihoods of those who depend so much on fragile lands.

> *We need to start measuring natural capital correctly and ensure that all decisions regarding its use are made on the basis of its real value to society. In this regard, we need to insist that Ministries of Finance reflect financial decisions on the basis of real costs and values.*[6]

Notes

1 Loss of carbon fixation from the atmosphere, associated with land degradation over the period, amounts to almost a thousand million tons. At a shadow price of US$50 per ton, the cost is almost US$50 billion. The cost of land degradation is at least an order of magnitude greater in terms of emissions to the atmosphere than through the impact of loss of soil organic carbon.

2 Some 16 per cent of the total land area shows improvement. Eighteen per cent of the improving area is cropland (20 per cent of the total croplands), 23 per cent is forest and 43 per cent rangeland.

3 There is a substantial body of evidence on the monetary values attached to biodiversity and ecosystems, and thus on the costs of their loss. A number of recent case studies and more general contributions have been received in reply to a call for evidence (see The Economics and Ecosystems and Biodiversity (TEEB) website, http://ec.europa.eu/ environment/nature/biodiversity/economics/index_en.htm, for a list of submissions and a synthesis report).

4 A particular effort is needed to assess indirect use values, especially those of regulating services, which are receiving increasing attention as a consequence of the Millennium Ecosystem Assessment (MEA). For carbon sequestration, substantial values have often been found, although they vary depending on the type of forest – for example, deciduous or coniferous – and their geographical location.

5 The on-site impacts of land degradation on productivity are easily masked due to use of additional inputs and adoption of improved technology and have led some to question the negative effects of desertification. The masking effect of improved technology can provide a false sense of security.

6 Ian Johnson, former vice-president of the World Bank in a Keynote Speech at Land Day 2, organized by the UNCCD secretariat on 5 June 2010 in Bonn.

References

Bai, Z., Dent, D., Olsson, L. and Schaepman, M. (2008) 'Global assessment of land degradation and improvement: 1. Identification by remote sensing', ISRIC Report 2008/01, ISRIC – World Soil Information, Wageningen

Birol, E., Kontoleon, A. and Smale. M. (2005) 'Farmer demand for agricultural biodiversity in Hungary's transition economy: A choice experiment approach', in M. Smale (ed) *Valuing Crop Genetic Biodiversity on Farms during Economic Change*, CAB International, Wallingford

Di Falco, S. and Perrings, C. (2005) 'Crop biodiversity, risk management and the implications of agricultural assistance', *Ecological Economics*, vol 55, no 4, pp459–466

Gardi C. and Jeffrey S. (2009) *Soil Biodiversity*, Joint Research Centre, European Commission, Luxembourg

GEF-STAP (Global Environment Facility Scientific and Technical Advisory Panel) (2010) 'New science, new opportunities for GEF-5 and beyond', report of the Scientific and Technical Advisory Panel to the 4th GEF Assembly, Punta Del Este, Uruguay

GEO3 Factsheet on Africa (2002) 'Past and present 1972 to 2002', United Nations Environment Programme, www.unep.org/geo/

Lal, R. (1998) 'Soil erosion impact on agronomic productivity and environment quality', *Critical Reviews in Plant Sciences*, vol 17, no 4, pp319–464

Millennium Ecosystem Assessment (2005) 'Millennium Ecosystem Assessment',

Ecosystems and Human Well-being: Desertification Synthesis, Island Press, Washington, DC

Pimentel, D., Wilson, C., McCullum, C., Huang, R., Dwen, P., Flack, J., Tran, Q., Saltman, T. and Cliff, B. (1997) 'Economic and environmental benefits of biodiversity', *Bioscience*, vol 47, no 11, pp747–757

Reich, P. F., Numbem, S. T., Almaraz, R.A. and Eswaran, H. (2001) 'Land resource stresses and desertification in Africa', *Agro-Science*, vol 2, no 2, pp1–10, http://ajol.info/index.php/as/article/view/1484

Schils, R., Kuikman, P., Liski, J., van Oijen, M., Smith, P., Webb, J., Alm, J., Somogyi, Z., van den Akker, J., Billett, M., Emmett, B., Evans, C., Lindner, M., Palosuo, T., Bellamy, P., Jandl, R. and Hiederer, R. (2008) *Review of Existing Information on the Interrelations Between Soil and Climate Change*, European Commission, Brussels

Srinivasan, T., Carey, S., Hallstein, E., Higgins, P., Kerr, A., Koteen, L., Smith, A., Watson, R., Harte, J. and Norgaard, R. (2008) 'The debt of nations and the distribution of ecological impacts from human activities', *Proceedings of the National Academy of Science of the United States of America*, vol 105, no 5, pp1768–1773

UNEP (2007) *Global Environmental Outlook: GEO 4*, United Nations Environment Programme, Nairobi

13

Vulnerable States

Jon Hutton

Introduction

Climate change poses significant risks and challenges to the security, stability and development capacities of many states around the world. The vulnerability of a state, defined as the degree to which it is likely to experience harm from exposure to climate change-related risks, will depend not only on the nature of climatic changes taking place within its territory, but also on the societal capacities to respond to the challenge.

Human populations can be significantly affected by climate change. Most changes come via the effects of climate change on infrastructure and ecosystems – the 'green infrastructure' of our planet. The degree of ecosystem change and degradation resulting from climate change is a key component of state vulnerability, as the many services provided by ecosystems are essential to human well-being in all countries (Reid et al, 2005).

Poor people often rely directly on such services for their livelihoods and basic needs, and the degradation of ecosystems from climate change threatens to exacerbate existing factors undermining development, such as food insecurity and disease. This is one of the reasons why developing countries are considered to be inherently more vulnerable to climate change, in addition to the fact that many of them experience significant financial, administrative and human capacity constraints to adaptation.

Changes in the climatic properties of the atmosphere can have direct impacts on ecosystem structure, composition and functioning. They can also increase the likelihood of natural disturbances, such as wildfires, pest outbreaks and encroachment of invasive alien species (IAS) into previously undisturbed habitats. These processes can have long-term implications for ecosystems, especially where they cross a critical threshold, leading to the transition of the ecosystem to a new state or even to its collapse. In many cases, the resulting changes reduce the capacity of the ecosystem to provide ecosystem services.

This chapter gives an overview of the vulnerability of ecosystems to climate change-induced risks in different regions of the world, and the significance of this vulnerability to the well-being of societies and communities. It focuses on four of the major risks that have been identified at the global level: drought, fire, flooding and sea-level rise. Interactions with other ecosystem change drivers that are not climate derived, but have the capacity to contribute to the vulnerability of a state's ecosystems to climate change, are also considered, as well as the potential for climate change to contribute to positive feedback loops within ecosystems.

Regional distribution of climate change-related risks

Forecasting the impacts of climate change on ecosystems in states around the world is not a simple task, since the temporal and spatial scales of exposure to changes in climatic factors, as well as the degree of resilience that will be exhibited by ecosystems, are insufficiently known (Fischlin et al, 2007). The situation is further compounded by the fact that many risks are a function of both climatic and non-climatic drivers, and differentiating between these two sets of drivers can be more or less difficult depending on the risk being considered. Nevertheless, observed impacts to date (Nelson, 2005; Walther, 2010), coupled with modelled scenarios, are helping to clarify what climate change may mean for the world's ecosystems. Both sources of information lead us to expect with high certainty that ecosystems will continue to undergo increasing change as a result of climate change within the next century (Fischlin et al, 2007). While it is expected that all regions of the world will experience tangible impacts, 'hot spots' can be identified and current projections show that risks from climate change will tend to disproportionally affect less developed regions of the world (Schneider et al, 2007).

Drought

The frequency and intensity of droughts and the extent of affected areas have been predicted to increase in a warmer world (Wetherald and Manabe, 2002; Kundzewicz et al, 2007). Drought is typically a common aspect of many regions' long-term hydrological regimes. Through greater occurrence and intensity, drought is projected to increase ecosystem vulnerability by depleting water stores, increasing fire hazards and exceeding the tolerance thresholds of certain types of vegetation to water stress, leading to a rising risk of desertification. Among the countries where dry spells are expected to increase, those that have ecosystems in arid and semi-arid zones, such as countries in the Sahel, Horn and southern regions of Africa, are areas of particular concern. This includes states with continental desert ecosystems, many of which are projected to experience more severe, persistent droughts and desertification risk (Lioubimtseva and Adams, 2004; Schwinning and Sala, 2004; Burke et al, 2006). Communities living in these regions are likely to be impacted through more frequent and severe water supply shortages, as well as the spread of uninhabitable areas as desertification progresses. There are concerns that increasing drought episodes could trigger famine and forced migration (GACGC, 2007).

However, state vulnerability to drought will not be confined to areas with arid ecosystems. Recent research (Malhi et al, 2009) has confirmed the concern that states harbouring tropical forest ecosystems could be at risk from increased drought incidence and intensity. A notable example is the Amazon rainforest, shared by Brazil, Bolivia, Colombia, Ecuador, French Guiana, Guyana, Peru, Suriname and Venezuela. Further, major wetland ecosystems, such as those in the delta regions of Pakistan, Bangladesh, India and China, are already being degraded by observed precipitation declines and droughts (Fischlin et al, 2007). This has serious implications for communities inhabiting these areas, many of which rely on the ecosystems for food and fuel resources. In Australia, prolonged periods of drought have greatly diminished water supplies in the southeastern regions, which, coupled with increasing population pressures, has led to the prediction that water insecurity will be a major issue for the area in coming decades.

Fire

In many ecosystems, rising temperatures and more frequent and intense droughts are already known to increase the likelihood of wildfires, both in size and frequency (Stocks et al, 1998; Podur et al, 2002; Brown et al, 2004). As climate change progresses, a higher prevalence of fire-related disturbances is likely (Gillet et al, 2004; Westerling et al, 2006). This is of global relevance, as fire prone vegetation types currently cover 40 per cent of the world's land surface (Chapin et al, 2002), being mostly located in tropical, subtropical and boreal regions (Harden et al, 2000; Bond et al, 2005). In eastern Canada, intensified wildfire regimes, driven partly by climate change, appear to be changing vegetation structure and composition (Lavoie and Sirois, 1998). Observations spanning the entire North American boreal region have found that total burned area from fires has increased by a factor of 2.5 between the 1960s and 1990s, while the area burned from human-ignited fires remained constant (Kasischke and Turetsky, 2006). Because boreal ecosystems are a significant carbon sink, helping to store up to one fifth of global terrestrial carbon, increased burning of these ecosystems has the potential to significantly contribute to global anthropogenic emissions.

Forest fires attributable to climate change are also becoming more common in tropical forest ecosystems such as those of the Amazon basin (Cochrane, 2003) and have had damaging effects on vegetation (Cochrane and Laurance, 2002; Haugaasen et al, 2003). This, coupled with other drivers such as El Niño-related droughts, deforestation, selective logging and forest fragmentation, has increased tropical forest susceptibility to fire in recent years (Fearnside, 2001; Nepstad et al, 2002; Cochrane, 2003), and the trend is expected to continue (Nepstad et al, 2004). In many regions of Asia, increasing intensity and spread of forest fires partially attributed to climate change have been observed over the past 20 years (Page et al, 2002; De Grandi et al, 2003; Goldammer et al, 2003; FFARF, 2004; Murdiyarso et al, 2004; Achard et al, 2005; Murdiyarso and Adininsih, 2007). Ecosystems particularly at risk include Siberian and Southeast

Asian peatlands: recent studies have found significant increases of Siberian peatland wildfires as a result of climate change (especially temperature increases) interacting with human-induced pressures (Cruz et al, 2007).

Forest dependent communities and settlements situated near forests can be seriously threatened by wildfires, both through direct risks to human health and damage to existing infrastructure, as well as through the diminution or elimination of forest resources and services.

Flooding

Floods are the most frequent natural disaster related to weather events (EM-DAT, 2006), and come in many forms, including river floods and floods associated with storm surges and coastal inundation. Flood magnitude and frequency are predicted to increase for most regions in the world for a variety of reasons (Parry et al, 2007). In some countries in Asia, the number of floods is expected to increase as a result of accelerated glacial melt. Increased rainfall intensity, especially in the summer monsoon months, may increase flood prone areas in temperate and tropical regions of Asia (Cruz et al, 2007). In Latin America, significant impacts from flooding associated with sea-level rise are expected in coastal areas for the 2050–2080 period. This is particularly relevant for mangrove ecosystems, as flooding is known to induce changes in mangrove distribution (Conde, 2001; Medina et al, 2001; Villamizar, 2004). Future risks from floods in coastal ecosystems are also likely to be significant in Africa. Recent assessments of potential flood risks that could arise by 2080 show that of the five regions most at risk from flooding in coastal and deltaic areas, three are from Africa (Warren et al, 2006). Apart from direct dangers to human security and infrastructures, there will be implications for communities dependent on aquaculture industries. Flooding is also known to further exacerbate the spread of diseases such as diarrhoea, cholera and malaria. In a warmer world, where flood magnitude and frequency are expected to increase, the contribution of flooding to human health risks will therefore be considerable (Few et al, 2004). Risks to communities in the future will be greatly determined by the extent to which populations reside in floodplains, the resilience of surrounding ecosystems and the strength of community flood response plans.

Sea-level rise

Global sea levels have risen at an average of 1.7mm/year during the 20th century (Church and White, 2006), mainly due to ocean thermal expansion, melting of glacial ice sheets and changes in terrestrial water storage. Recent research (Pfeffer et al, 2008; Lowe et al, 2009) has indicated that total sea-level rise could amount to one metre or more by the end of this century. This has considerable implications for regions of the world with a large extent of coastal and low-lying ecosystems, such as mangroves and coastal wetlands. Sea-level rise is expected to negatively impact mangrove forests in coastal areas of Latin America (Kovacs, 1999; Meagan et al, 2004). In Asia, a 1m rise in sea level will mean a 2500km^2 loss of mangroves, including the Sundurbans mangrove ecosystem, which is shared by India and Bangladesh, and is a global priority zone for biodiversity (Loucks et al, 2010). In Africa, increased flooding risks along coastal and deltaic areas will be a direct consequence of sea-level rise. The most vulnerable states are those situated on small low lying islands, and some of them are in danger of being completely submerged. In the Maldives, up to 80 per cent of the country's area is 1m or less above sea level. For a country that depends almost exclusively on coastal and marine resources for its livelihoods and economy (Emerton et al, 2009), sea-level rise is a danger threatening the country's very existence. The contribution of sea-level rise to international and intra-national human migration and displacement is likely to exert pressure on states, as well as on the capacities of areas into which people move to support communities and livelihoods.

The many dimensions of climate change risk

Climate change can affect ecosystems in different ways, with impacts often being a consequence of several factors working in combination. This is particularly true for climate change impacts in forest ecosystems. For example, a forest may undergo damage from extreme wind events that result in branch breaking, crown loss, trunk breakage or complete stand destruction (Easterling et al, 2007). Further damage to the ecosystem can then result from wildfire and insect outbreaks (Fleming et al, 2002; Nabuurs et al, 2002). Some climate-related risks require other types of climate impacts to establish suitable conditions in an ecosystem before they can take place. For example, wildfires are more likely to occur if a forest ecosystem has been exposed to prolonged drought, which increases the flammability of a forest (Nepstad et al, 2004). As anthropogenic interference with the climate continues to alter the atmospheric properties of the planet, the complex interactions between different climate risks are likely to further evolve. Efforts to reduce state vulnerability will therefore need to consider and anticipate the interactions between climate-related risks, as well as their continuously evolving nature.

Interactions with other drivers of ecosystem change

Although climate change is an important driver of ecosystem change and degradation, it is not the only one. Currently, many of the major threats to ecosystems can be attributed to changes in land use, resource extraction and other related processes (such as nitrogen deposition, pollution and human-induced fires). By contributing to ecosystem degradation, these pressures have the capacity to reduce resilience and increase vulnerability to climate change. For example, land-use change can lead to habitat fragmentation, which can hinder the ability of species to migrate, thereby reducing their capacity to naturally adapt through geographical range shifts (Fischlin et al, 2007). Ecosystem degradation through biodiversity loss follows, as does decreased capacity of the ecosystem to withstand external pressures such as climate change. A further example is where human-induced fires have transformed certain types of forests to shrublands and grasslands, which are more flammable (Ogden et al, 1998) and therefore at higher risk from climate-related fire outbreaks.

Another link between non-climate-related anthropogenic pressures and ecosystem vulnerability is that many of these pressures result in the depletion of ecosystem carbon stocks, thus contributing to further climate change. In southeast Asia, for example, human activities have changed fire regimes in a way that have turned certain ecosystems into a significant source of carbon emissions. In Indonesia, the burning or draining of carbon-rich peatlands to clear land for agriculture and palm oil plantations in recent decades has released significant amounts of carbon into the Earth's atmosphere, which accounted for roughly 10 per cent of global carbon emissions in 2007.

Positive feedbacks and critical thresholds

The unprecedented rate and extent of projected climate change, coupled with other rising human pressures, has led to the fear that climate change could lead to ecosystems reaching critical thresholds, or 'tipping points' (abrupt ecosystem shifts towards novel states which are currently poorly understood) (Fischlin et al, 2007; Secretariat of the CBD, 2010). The likelihood of reaching such thresholds may be increased by the existence of 'positive feedbacks' between ecosystem responses to climate change and climate change itself.

A recent study of the risk of a climate change-induced dieback of the Amazon forest found that dry-season water stress is likely to intensify in the eastern Amazon over the 21st century, thereby increasing the ecosystem's vulnerability to fires (Malhi et al, 2009). This could lead to a critical threshold being reached when combined with overall rising global temperatures and human-induced pressures such as deforestation, logging and fragmentation, beyond which ecosystem structure and function will irreversibly change. Another area of high concern in terms of positive feedback systems is permafrost thawing in the Arctic region. As the permanently frozen subsoil typical of polar regions thaws, it releases organic carbon into the atmosphere in the form of methane, a potent GHG. Observations of recent permafrost thawing have concluded that earlier estimates of methane release from permafrost carbon stocks were low, and that current rates of release, along with changes in albedo (the capacity of a surface to reflect sunlight), through the Arctic region will make a further contribution to climatic warming (Camill, 2005; Lelieveld, 2006; Walter et al, 2006; Zimov et al, 2006).

Conclusions

State vulnerability to the risks from climate change operates on many levels. The inextricable link between a country's ecosystems and community well-being make the issue of ecosystem vulnerability a human development problem. Although some uncertainties remain regarding the exact nature, degree and timing of ecosystems' response to climate change, it is clear that ecosystems will be significantly affected.

Evidence to date has shown that the four major global climate change risks addressed in this chapter are already presenting serious challenges for ecosystems and the development pathways of countries. These challenges are in many cases exacerbated by the presence of other human pressures, which intensify impacts to ecosystems and, subsequently, to communities. The combination of climate and non-climate-related pressures also act to decrease the resilience of many ecosystems, thereby curtailing their ability to continue providing essential services under future climate change.

The positive news is that there are many actions that can be taken now to significantly reduce the extent of state and ecosystem vulnerability to climate change in the future. These include reducing human pressures such as deforestation that, in addition to contributing to ecosystem destruction, further propagate global GHG emissions. A better understanding of processes that could induce tipping points or lead to critical thresholds being reached is needed if we are to have a chance of avoiding the most extreme ecosystem responses to climate change. When establishing societal response plans, a clear effort needs to be made to avoid considering in isolation the processes leading to ecosystem vulnerability. Finally, international resources and efforts need to be directed to areas of the world where risks from climate change are high, and societal response capacities are low.

References

Achard, F., Stibig, H. J., Laestadius, L., Roshchanka, V., Yaroshenko, A. and Aksenov, D. (eds) (2005) *Identification of 'Hotspot Areas' of Forest Cover Changes in Boreal Eurasia*, Office for Official Publication of the European Communities, Luxembourg, www-tem.jrc.it//PDF_publis/2005/Achard&al_HotSpot-Boreal-Eurasia.pdf

Bond, W. J., Woodward F. I. and Midgley, G. F. (2005) 'The global distribution of ecosystems in a world without fire', *New Phytologist*, vol 165, no 2, pp525–537

Brown, T. J., Hall B. L. and Westerling, A. L. (2004) 'The impact of twenty-first century climate change on wildland fire danger in the western United States: An applications perspective', *Climatic Change*, vol 62, nos 1–3, pp365–388

Burke, E. J., Brown S. J. and Christidis, N. (2006) 'Modeling the recent evolution of global drought and projections for the 21st century with the Hadley Centre climate model', *Journal of Hydrometeorolly*, vol 7, no 5, pp1113–1125

Camill, P. (2005) 'Permafrost thaw accelerates in boreal peatlands during late-20th century climate warming', *Climatic Change*, vol 68, nos 1–2, pp135–152

Chapin, F. S., Matson, P. A. and Mooney, H. A. (2002) *Principles of Terrestrial Ecosystem Ecology*, Springer, New York

Church, J. A. and White, N. J. (2006) 'A 20th century acceleration in global sea level rise', *Geophysical Research Letters*, vol 33, L01602, doi:10.1029/2005GL024826

Cochrane, M. A. (2003) 'Fire science for rainforests', *Nature*, vol 421, no 6926, pp913–919

Cochrane, M. A. and Laurance, W. F. (2002) 'Fire as a large-scale edge effect in Amazonian forests', *Journal of Tropical Ecology*, vol 18, no 3, pp311–325

Conde, J. E. (2001) 'The Orinoco River Delta, Venezuela', in U. Seeliger and B. Kjerfve (eds) *Coastal Marine Ecosystems of Latin America*, Springer-Verlag, Berlin, pp61–70

Cruz, R. V., Harasawa, H., Lal, M., Wu, S., Anokhin, Y., Punsalmaa, B., Honda, Y., Jafari, M., Li, C. and Huu Ninh, N. (2007) 'Asia', in M. L. Parry, O. F. Canziani,

J. P. Palutikof, P. J. van der Linden and C. E. Hanson (eds) *Climate Change 2007: Impacts, Adaptation and Vulnerability. Contribution of Working Group II to the Fourth Assessment Report of the Intergovernmental Panel on Climate Change*, Cambridge University Press, Cambridge, pp469–506

De Grandi, G. F., Pauste, Y., Achard, F. and Mollicone, D. (2003) *The GBFM Radar Mosaic of the Eurasian Taiga: Groundwork for the Bio-physical Characterization of an Ecosystem with Relevance to Global Change Studies*, International Geoscience and Remote Sensing Symposium, Toulouse

Easterling, W. E., Aggarwal, P. K., Batima, P., Brander, K. M., Erda, L., Howden, S. M., Kirilenko, A., Morton, J., Soussana, J. F., Schmidhuber, J. and Tubiello, F. N. (2007) 'Food, fibre and forest products', in M. L. Parry, O. F. Canziani, J. P. Palutikof, P. J. van der Linden and C. E. Hanson (eds) *Climate Change 2007: Impacts, Adaptation and Vulnerability. Contribution of Working Group II to the Fourth Assessment Report of the Intergovernmental Panel on Climate Change*, Cambridge University Press, Cambridge, pp273–313

EM-DAT (2006) *The OFDA/CRED International Disaster Database*, www.em-dat.net

Emerton, L., Baig, S. and Saleem, M. (2009) *Valuing Biodiversity: The Economic Case for Biodiversity Conservation in the Maldives*, AEC Project, Ministry of Housing, Transport and Environment, Government of Maldives and United Nations Development Programme, Maldives

Fearnside, P. M. (2001) 'Status of South American natural ecosystems', in S. Levin (ed) *Encyclopedia of Biodiversity: Volume 5*, Academic Press, pp345–359

Few R., Ahern, M., Matthies, F. and Kovats, S. (2004) 'Floods, health and climate change: A strategic review', Working Paper no 63, Tyndall Centre for Climate Change Research, University of East Anglia, Norwich

FFARF (Federal Forest Agency of the Russian Federation) (2004) *State Forest Account of Russia for the Year 2003*, Russian Scientific Research Centre, Lesresurs, Moscow

Fischlin, A., Midgley, G. F., Price, J. T., Leemans, R., Gopal, B., Turley, C., Rounsevell, M. D. A., Dube, O. P., Tarazona, J. and Velichko, A. A. (2007) 'Ecosystems, their properties, goods, and services', in M. L. Parry, O. F. Canziani, J. P. Palutikof, P. J. van der Linden and C. E. Hanson (eds) *Climate Change 2007: Impacts, Adaptation and Vulnerability. Contribution of Working Group II to the Fourth Assessment Report of the Intergovernmental Panel on Climate Change*, Cambridge University Press, Cambridge, pp211–272

Fleming, R. A., Candau J. N. and McAlpine, R. S. (2002) 'Landscape-scale analysis of interactions between insect defoliation and forest fire in Central Canada', *Climatic Change*, vol 55, nos 1–2, pp251–272

GACGC (German Advisory Council on Global Change) (2007) *World in Transition: Climate Change as a Security Risk*, Earthscan, London

Gillett, N. P., Weaver, A. J., Zwiers, F. W. and Flannigan, M. D. (2004) 'Detecting the effect of climate change on Canadian forest fires', *Geophysical Research Letters*, vol 31, L18211

Goldammer, J. G., Sukhinin, A. I. and Csiszar, I. (2003) 'The current fire situation in the Russian Federation', *International Forest Fire News*, vol 29, pp89–111

Harden, J. W., Trumbore, S. E., Stocks, B. J., Hirsch, A., Gower, S. T., O'Neill, K. P. and Kasischke, E. S. (2000). 'The role of fire in the boreal carbon budget', *Global Change Biology*, vol 6, no 2, pp174–184

Haugaasen, T., Barlow, J. and Peres, C. A. (2003) 'Surface wildfires in central Amazonia: Short-term impact on forest structure and carbon loss', *Forest Ecology and Management*, vol 179, pp321–331

Kasischke, E. S. and Turetsky, M. R. (2006) 'Recent changes in the fire regime across the North American boreal region: Spatial and temporal patterns of burning across Canada and Alaska', *Geophysical Research Letters*, vol 33, L09703

Kovacs, J. M. (1999) 'Assessing mangrove use at the local scale', *Landscape and Urban Planning*, vol 43, no 4, pp201–208

Kundzewicz, Z. W., Mata, L. J., Arnell, N. W., Döll, P., Kabat, P., Jiménez, B., Miller, K. A., Oki, T., Sen, Z. and Shiklomanov, I. A. (2007) 'Freshwater resources and their management', in M. L. Parry, O. F. Canziani, J. P. Palutikof, P. J. van der Linden and C. E. Hanson (eds) *Climate Change 2007: Impacts, Adaptation and Vulnerability. Contribution of Working Group II to the Fourth Assessment Report of the Intergovernmental Panel on Climate Change,* Cambridge University Press, Cambridge, pp173–210

Lavoie, L. and Sirois, L. (1998) 'Vegetation changes caused by recent fires in the northern boreal forest of eastern Canada', *Journal of Vegetation Science*, vol 9, no 4, pp483–492

Lelieveld, J. (2006) 'Climate change: A nasty surprise in the greenhouse', *Nature*, vol 443, no 7110, pp405–406

Lioubimtseva, E. and Adams, J. M. (2004) 'Possible implications of increased carbon dioxide levels and climate change for desert ecosystems', *Environmental Management*, vol 33, supp 1, ppS388–S404

Loucks, C., Barber-Meyer, S., Hossain, A. A., Barlow, A. and Chowdhury, R. M. (2010) 'Sea level rise and tigers: Predicted impacts to Bangladesh's Sundarbans mangroves', *Climatic Change*, vol 98, nos 1–2, pp291–298

Lowe, J. A., Howard, T., Pardaens, A., Tinker, J., Jenkins, G., Ridley, J., Leake, J., Holt, J., Wakelin, S., Wolf, J., Horsburgh, K., Reeder, T., Milne, G., Bradley, S. and Dye, S. (2009) *UK Climate Projections Science Report: Marine and Coastal Projections*, Met Office, Hadley Centre, Exeter

Malhi, Y., Aragão, L. E. O. C., Galbraith, D., Huntingford, C., Fisher, R., Zelazowski, P., Sitch, S., McSweeney, C. and Meir, P. (2009) 'Exploring the likelihood and mechanism of a climate-change-induced dieback of the Amazon rainforest', *PNAS*, vol 106, no 49, pp20610–20615

Meagan, E., Adina G. and Herrera-Silveira, J. A. (2004) 'Tracing organic matter sources and carbon burial in mangrove sediments over the past 160 years', *Estuarine, Coastal and Shelf Science*, vol 61, no 2, pp211–227

Medina, E., Fonseca, H., Barboza, F. and Francisco, M. (2001) 'Natural and man-induced changes in a tidal channel mangrove system under tropical semiarid climate at the entrance of the Maracaibo Lake (western Venezuela)', *Wetlands Ecology Management*, vol 9, no 3, pp243–253

Millennium Environmental Assessment (2005) *Ecosystems and Human Well-being: Synthesis*, Island Press, Washington, DC

Murdiyarso, D. and Adiningsih, E. (2007) 'Climatic anomalies, Indonesian vegetation fires and terrestrial carbon emissions', *Mitigation and Adaptation Strategies for Global Change*, vol 12, no 1, pp101–112

Murdiyarso, D., Lebel, L., Gintings, A. N., Tampubolon, S. M. H., Heil, A. and Wasson, M. (2004) 'Policy responses to complex environmental problems: Insights from a science-policy activity on transboundary haze from vegetation fires in Southeast Asia', *Agriculture, Ecosystems & Environment*, vol 104, no 1, pp47–56

Nabuurs, G. J., Pussinen, A., Karjalainen, T., Erhard, M. and Kramer, K. (2002) 'Stemwood volume increment changes in European forests due to climate change – a simulation study with the EFISCEN model', *Global Change Biology*, vol 8, no 4, pp304–316

Nelson, G. C. (2005) 'Drivers of ecosystem change: Summary chapter', in R. Hassan, R. Scholes and N. Ash (eds) *Ecosystems and Human Well-being: Volume 1: Current State and Trends*, Island Press, Washington, DC, pp73–76

Nepstad, D., McGrath, D., Alencar, A., Barros, A. C., Carvalho, G., Santilli, M. and Vera Diaz, M. D. C. (2002) 'Frontier governance in Amazonia', *Science*, vol 295, no 5555, pp629–631

Nepstad, D., Lefebvre, P., da Silva, U. L., Tomasella, J., Schlesinger, P., Solórzano, L., Moutinho, P., Ray, D. and Benito, J. G. (2004) 'Amazon drought and its implications for forest flammability and tree growth: A basin-wide analysis', *Global Change Biology*, vol 10, no 5, pp704–717

Ogden, J., Basher, L. and McGlone, M. (1998) 'Fire, forest regeneration and links with early human habitation: Evidence from New Zealand', *Annals of Botany*, vol 81, no 6, pp687–696

Page, S. E., Siegert, F., Rieley, J. O., Boehm, H. D. V., Jaya, A. and Limin, S. (2002) 'The amount of carbon released from peat and forest fires in Indonesia during 1997', *Nature*, vol 420, no 6911, pp61–65

Parry, M. L., Canziani, O. F., Palutikof, J. P. (2007) 'Technical summary', in M. L. Parry, O. F. Canziani, J. P. Palutikof, P. J. van der Linden and C. E. Hanson (eds) *Climate Change 2007: Impacts, Adaptation and Vulnerability. Contribution of Working Group II to the Fourth Assessment Report of the Intergovernmental Panel on Climate Change*, Cambridge University Press, Cambridge, pp23–78

Pfeffer, W. T., Harper, J. T. and O'Neel, S. (2008) 'Kinematic constraints on glacier contributions to 21st-century sea-level rise', *Science*, vol 321, no 5894, pp1340–1343

Podur, J., Martell, D. L. and Knight, K. (2002) 'Statistical quality control analysis of forest fire activity in Canada', *Canadian Journal of Forest Research*, vol 32, no 2, pp195–205

Schneider, S. H., Semenov, S., Patwardhan, A., Burton, I., Magadza, C. H. D., Oppenheimer, M., Pittock, A. B., Rahman, A., Smith, J. B., Suarez A. and Yamin, F. (2007) 'Assessing key vulnerabilities and the risk from climate change', in M. L. Parry, O. F. Canziani, J. P. Palutikof, P. J. van der Linden and C. E. Hanson (eds) *Climate Change 2007: Impacts, Adaptation and Vulnerability. Contribution of Working Group II to the Fourth Assessment Report of the Intergovernmental Panel on Climate Change*, Cambridge University Press, Cambridge, pp779–810

Schwinning, S. and Sala, O. E. (2004) 'Hierarchy of responses to resource pulses in arid and semi-arid ecosystems', *Oecologia*, vol 141, no 2, pp211–220

Secretariat of the CBD (Convention on Biological Diversity) (2010) *Global Biodiversity Outlook 3*, Convention on Biological Diversity, Montreal

Stocks, B. J., Fosberg, M. A., Lynham, T. J., Mearns, L., Wotton, B. M., Yang, Q., Jin, J. Z., Lawrence, K., Hartley, G. R., Mason, J. A. and McKenney, D. W. (1998) 'Climate change and forest fire potential in Russian and Canadian boreal forests', *Climatic Change*, vol 38, no 1, pp1–13

Villamizar, A. (2004) 'Informe técnico de denuncia sobre desastre ecológico en el Desparramadero de Hueque', Edo. Falcón, Presentado ante La Fiscalía General y la Defensoría del Pueblo, Venezuela

Walter, K. M., Zimov, S. A., Chanton, J. P., Verbyla, D. and Chapin, F. S. (2006) 'Methane bubbling from Siberian thaw lakes as a positive feedback to climate warming', *Nature*, vol 443, no 7107, pp71–75

Walther, G. R. (2010) 'Community and ecosystem responses to recent climate change', *Philosophical Transactions of the Royal Society B: Biological Sciences*, vol 365, no 1549, pp2019–2024

Warren, R., Arnell, N., Nicholls, R., Levy, P. E and Price, J. (2006) 'Understanding the regional impacts of climate change: Research report prepared for the Stern Review', Tyndall Centre Working Paper no 90, Tyndall Centre, Norwich

Westerling, A. L., Hidalgo, H. G., Cayan, D. R. and Swetnam, T. W. (2006) 'Warming and earlier spring increase western US forest wildfire activity', *Science*, vol 313, no 5789, pp940–943

Wetherald, R. T. and Manabe, S. (2002) 'Simulation of hydrologic changes associated with global warming', *Journal of Geophysical Research*, vol 107, no D19, pp4379–4393

Zimov, S. A., Schuur, E. A. G. and Chapin, F. S. (2006) 'Permafrost and the global carbon budget', *Science*, vol 312, no 5780, pp1612–1613

14

Permafrost Dynamics and Global Climate Change

N. A. Shpolianskaya

Climate and permafrost are the two phenomena that form the basis of many of the natural features in the Arctic regions. The key problem in the permafrost area (cryolithozone) is the melting of the ice, which occurs broadly in the northern and boreal zones of the Earth. The frozen grounds, which have high strength and resistance under negative temperature, lose these properties when temperatures rise and the ice melts.

Climate

The cryolithozone, being a climate derivative, occurred when the climate became consistently cold more than 3 million years ago. However, the fluctuating nature of climate lapses (Figure 14.1 and Figure 14.2) conditioned repeated changes in cryolithozone conditions over the time of its existence: permafrost line, temperature and thickness, ice content and cryogenic process activity were changing.

The most noticeable climate changes occur in periods of approximately 40,000–45,000 years (Figures 14.1, 14.2 and 14.3). The cold period of the late Pleistocene (18,000–20,000 years ago) saw atmospheric temperatures of 7–10°C lower than current levels; and the peak of the warm Holocene period (5000–8000 years ago) resulted in atmospheric temperatures 2–2.5°C higher than present-day (Figure 14.3), when woodlands advanced to the north up to 68–70° of northern latitude. During latter periods, short temperature fluctuations were traced. Figure 14.4 demonstrates atmospheric temperature changes in historic time – for the past 4000 years and for 500 years. The middle of the first millennium AD was the 'historic ice period' that at the end of the first millennium was replaced by the warm 'Viking age' (900–1300 AD). In the 16th to 18th centuries the 'Minor Ice Age' occurred when the temperature dropped by 1.5–2°C, sea ice area expanded; and mountain glaciers enlarged. From the end of the 19th century, a new warming began, which was recorded with instrumental observations. This ascending branch is complicated with smaller waves, both in range and duration (approximately 30 years) (Figure 14.2). These waves showed warming in 1879, cooling in 1902–1905, warming with maximum temperature in 1930s ('Arctic warming era'), cooling in 1960–1970s, and finally the present warming of the 1990s. This provides evidence of cyclical fluctuations in temperature conditions.

There is geologic evidence of the occurrence of long and short-period cycles through geological history. It is identified in particular in Pre-Cambrian lacustrine-glacial varve clays (680 million years ago) in Australia and in Permian deposits (230–280 million years ago) in North America. It appears from this that temperature fluctuation is a natural characteristic of climate evolution. Most probably it will continue in the future. Presumably the whole variety of climatic cycles will continue as well. Extrapolating climatic cycles of the past 10,000 years to make future predictions (Figure 14.3), we can expect cooling within long-period cycles that in 15,000–20,000 years will lead to ice ages comparable with early (around 70,000 years ago) and late (18,000–20,000 years ago) Valdai. Shorter cycles with small ranges of temperature fluctuations will not cause noticeable rise of atmospheric temperature in future. We can assume there will not be significant climate changes due to short fluctuation periods in the next 3000 years.

Permafrost

The response of the cryolithozone to climate change relates to the penetration of atmospheric temperature fluctuations into rocks and further changes in cryolithozone temperature patterns. According to Fourier's laws, such penetration happens with a time lag that increases with depth and directly correlates to the period duration and fluctuation range. With period and range increase, the penetration depth increases as well (Tables 14.1 and 14.2) (Shpolianskaya, 2001, 2008).

Table 14.1 *Dependence of penetration depth (Z) of short-period temperature fluctuations on the period duration and fluctuation range*

Fluctuation period, T years	3		5		6		10		12		25		35	
A_o°C	2	3	2	3	2	3	2	3	2	3	2	3	2	3
Z,M	10	11	13	14	14	16	18	21	20	22	29	31	34	37

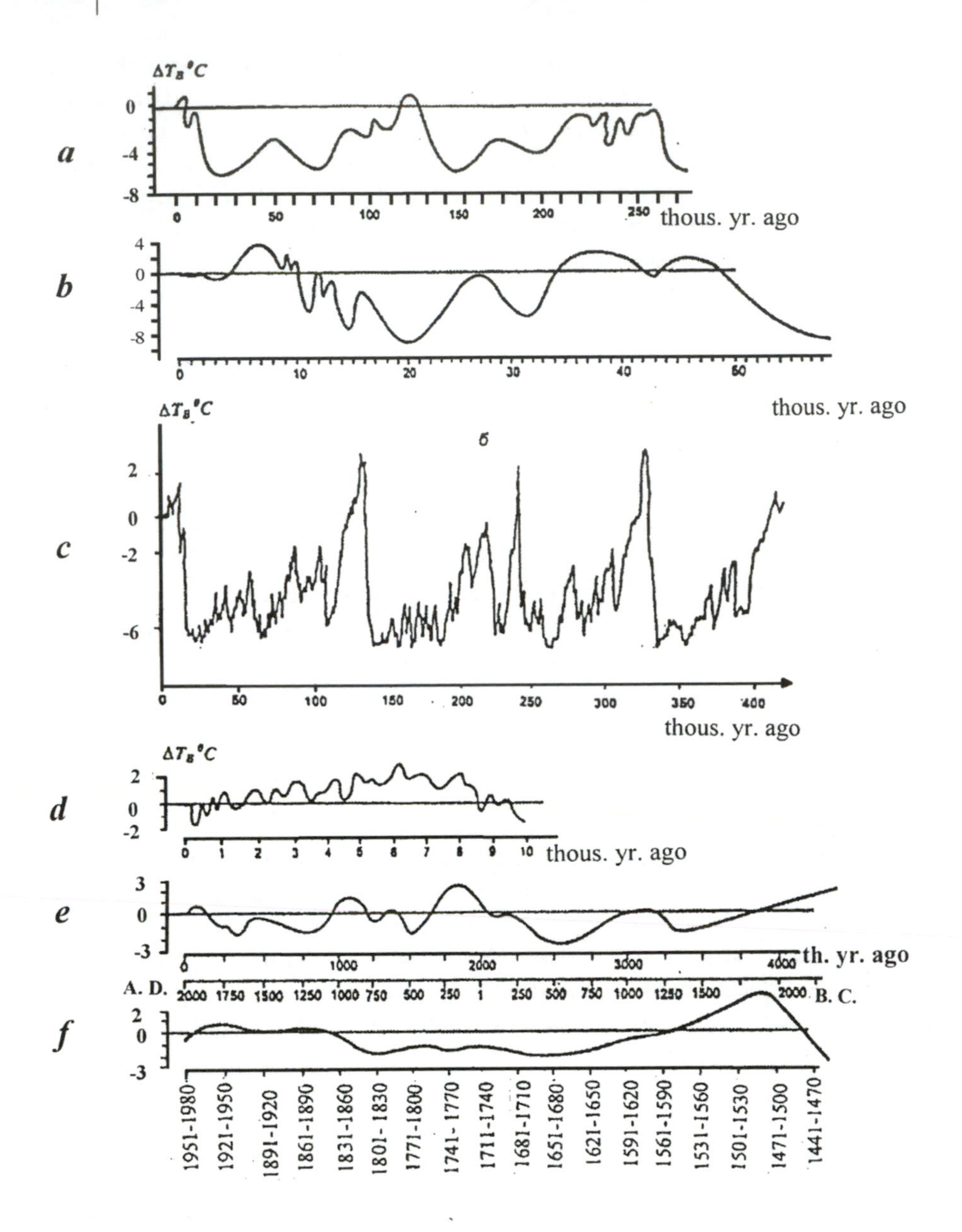

Figure 14.1 *Deviations of the past temperatures from present-day values*

Source: Shpolianskaya (2008)

Note: a – middle neo-Pleistocene-Holocene, North Atlantic; b – late neo-Pleistocene-Holocene, West Siberia; c – middle neo-Pleistocene-Holocene, Antarctic, Vostok; d – Holocene, the European part of Russia; e – the last 4000 years, the European part of Russia; e – for the last 500 years (average 30 year variations).

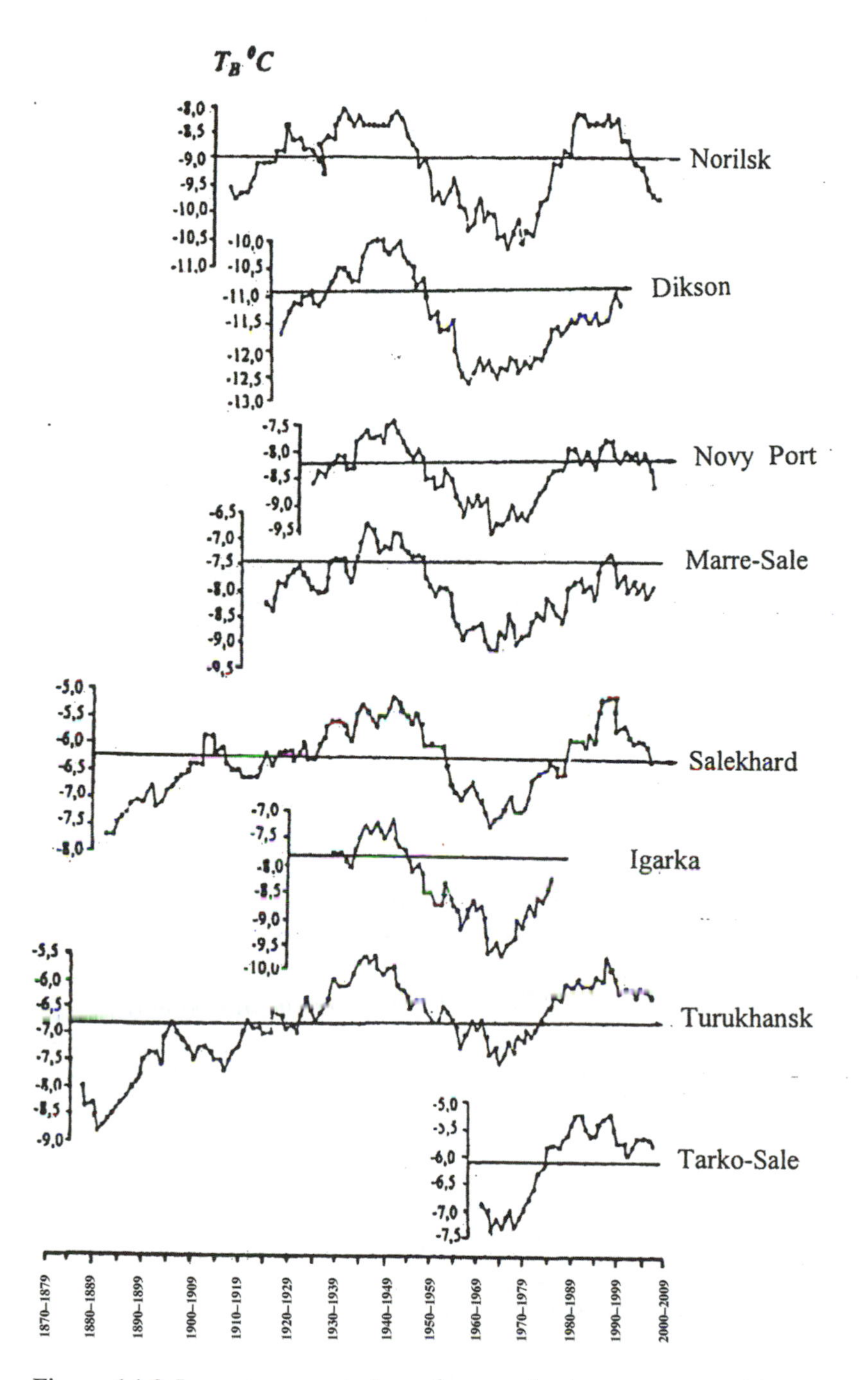

Figure 14.2 *Long-term variation of atmospheric temperature (T_B) per meteorological stations of West Siberia (moving average ten-year variations) for the period from the beginning of observations*

Source: Shpolianskaya (2008)

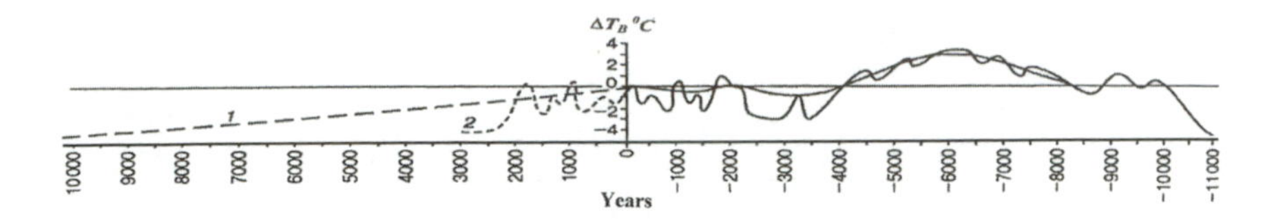

Figure 14.3 *Cumulative curve of atmospheric temperature deviations from present-day values for the last 10,000 years and forecast of its future variation*

Source: Shpolianskaya (2001)
Note: 1 – Assumed downward branch of temperature fluctuations within 45,000-year cycle; 2 – assumed short-period temperature fluctuations (as their continuation in historical period)

Table 14.2 *Dependence of penetration depth (Z) of average and long-period temperature fluctuations on the period duration (τ) and fluctuation range*

Fluctuation period, τ, years	100		500			1000			2000		3000			5000		
A_o°C	5	8	2	5	8	2	5	8	2	8	2	5	8	2	5	8
Z,M	74	83	127	165	185	180	234	263	331	371	312	405	455	400	523	586

Data from Tables 14.1 and 14.2 show that thermal waves that last for 5000, and even 3000, years penetrate deeper. For example, 4000–8000 years ago, climatic optimum warming in West Siberia with a range of 2–2.5°C changed the temperature field of the cryolithozone all the way down to 400m. Due to this, the permafrost thawed down to 200m to the south of the Polar Circle. Smaller circles of temperature fluctuations had a lesser impact if they were shorter and had a smaller range. About 2000–4000 years ago, cooling of about 2°C (Figure 14.6) did not penetrate lower than 130–180m. New frost penetration (down to 80–100m) into previously thawed formations occurred; however, it did not reach the level of the lower frozen layer that became connate. Frozen formation response to Holocene warming depended on its basic temperature (*T*). Thus, to the north of 68° of north latitude (in tundra) under low negative temperatures, frozen formations did not thaw from above. To the south, between 66° and 68° of north latitude (in forest tundra), under higher temperature, there were coexisting areas of unthawed frozen formations and areas partially thawed from above. To the south from 66° of north latitude (in forest area) under the temperature of about 0°C, frozen formations thawed from above everywhere. Therefore, the cryolithozone of West Siberia currently has four areas stretching from the north to the south (Figure 14.4). (Irpotin et al, 2008)

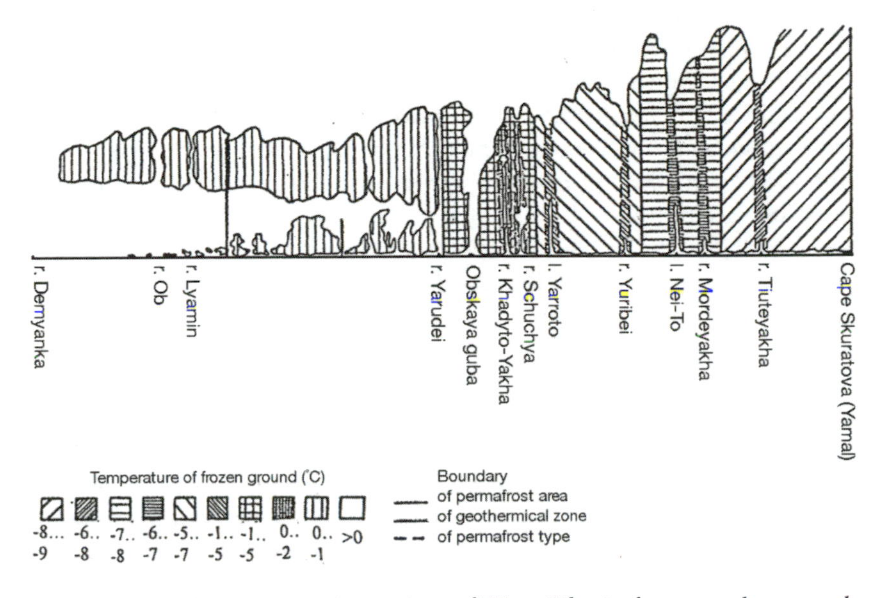

Figure 14.4 *Diagrammatic section of West Siberia from north to south and rock temperature*

Note: Area with from –9°C to –5°C, where permanent frost did not thaw. Area with *T* from –1°C to –5°C, where newly frozen formations merged with lower connate layer of frozen earth. Area with from 0°C to –2°C, where no merge of newly frozen formations and lower connate layer of frozen earth occurred and two-layer frozen formation developed. Area with *T*>0°C, where no new thawing emerged and only lower connate frozen formation was preserved.
Source: Shpolianskaya (2008)

Current changes in permafrost

Current changes in permafrost are conditioned by short-period climate variations (Figure 14.2). Short-period temperature fluctuations do not penetrate deep into subsoil rocks and affect the first tens of metres only. However, it is here where most human economic activity is concentrated. Therefore, forecasts for ground temperature variations resulting from short-period temperature fluctuations is very important. This task is complicated by the fact that warming started in the late 1970s and continues now, a fact that is considered by many authors to be human-caused via greenhouse gas (GHG) emissions. The warming is expected to be long and disrupt cyclical climate variations.

Permafrost response to present-day warming is well traced based on long-term monitoring investigations performed by A. V. Pavlov (2002, 2008). Pavlov carried out measurements of air and ground temperatures at a depth of 3m and 5m. In the northern part of Europe, at Pechora river (Bolvansky observing station), an average annual atmospheric temperature of –4.7°C and solid permafrost temperature variation at a depth of 9m did not correlate to atmospheric temperature variation during the period 1984–2001. There is practically no temperature variation. Atmospheric temperature fluctuations attenuate at a depth of 9m. While shifting to the east (to Vorkuta), which has a lower atmospheric temperature (–6.5°C), soil response to warming is more obvious. From 1950 to 1996, soil temperatures increased by 1.5°C at a depth of 3m, and by 0.5°C at a depth of 10m. In western Yamal in Siberia (Marre-Sale observation station), the average annual atmospheric temperature of -8°C and continuous permafrost soil temperature apparently respond to atmospheric temperature variations.

During 1978–1995, when atmospheric temperature increased by 3.2°C, soil temperature at a depth of 10m increased by 0.9°C on average. In this case, the terrain with low soil temperature (–5°C to –7°C) experienced an increase of 2.2°C at a depth of 3m and 1.3°C at a depth of 10m on average. However, within the terrains that had a temperature of –3°C to –4°C, soil temperature increased by 0.7°C only at a depth of 10m. In the next period (1996–2000), atmospheric temperature drop was followed by soil temperature decrease at a depth of 10m by 0.6°C. At Nadym observation station (average atmospheric temperature –5.9°C and fragmented permafrost), soil temperature at a depth of 10m increased by 0.2–0.8°C; and in hilly peat lands, by 0.4°C. At Urengoy oil and gas field, which has an atmospheric temperature of –9°C and continuous permafrost, soil temperature at a depth of 10m increased by 1°C on average during 1975–1999. In central Yakutia (Yakustk and at nearby Chabyda observation station), which has an average annual atmospheric temperature of –10.2°C and continuous permafrost, soil response to atmospheric temperature rise was insignificant during 1965–1999. Soil temperature fluctuations attenuate rapidly at a depth of 3–10m.

These observations demonstrate that the temperature of upper soil layers follows atmospheric temperature variations rising and dropping together year by. However, this relationship is also affected by local terrain features. Partial correlations with short-period fluctuations are also revealed. Directional soil temperature increase that should have been observed under human-caused warming is not traced.

At the same observation stations, Pavlov performed monitoring investigations of seasonal freezing and thawing depth variations, the layer where, due to maximum temperature gradients, the most active cryogenic processes take place. The key finding of the observations is a weak response to present-day climate warming. Long-term trends of seasonal freezing and thawing depth variations have both positive and negative values. Mostly they depend on specific lithologic conditions, soil moisture content and nature of land cover (snow and vegetation).

Cryogenic processes

Cryogenic processes occurring mostly in seasonal, active, freezing/thawing layers appear as specific cryogenic landforms. Diversity of the latter is formed by a few processes: soil cracking resulting from frost damage conditioned by temperature gradients forms 'polygonal terrain' with large ice veins along the cracks; soil cracking resulting from the moisture gradient and forming 'micro-polygonal terrain' not accompanied with subsurface ice formations; and soil heaving forming 'hilly terrain' (hillocks of swelling with an ice core). Cryogenic terrain was formed during the whole permafrost lifetime; however, its activity and development direction changed along with climatic period changes. Therefore, there is an 'upward development' model corresponding primarily to cold periods of stable frost accumulation; and 'downward development' corresponding to warming and permafrost degradation periods. Funnel types of topography – thermokarst (sinks, lakes) and soil flow – solifluction are brought about by ice thawing, thus rendering the territory not suitable for use.

Present-day warming promotes development of downward forms of cryogenic terrain. For example, based on observations in West Siberia (Kirpotin et al, 2008), there is a stable thermokarst lake growth trend (their area has increased by 12 per cent). However, the same observations demonstrate that, under common warming, cryogenic processes can have alternate trends that relate to corrective actions of the natural environment. Thus, thermokarst lake growth detected to the north of 67° north latitude, in the area of continuous permafrost, alternates with the weed population of lakes and their transformation into swamps (khasyrei). This is a typical situation. Thermokarst is developed actively in areas with subsurface ice (thawing ice produces water and forms lakes). Thawing of permafrost with low ice content does not produce lakes. Along 67° of north latitude is the southern border of thick subsurface ice deposits that have survived from the Pleistocene (border of the region where permafrost did not thaw in the Holocene period). Thinning of this ice intensified during the present-day warming preventing thermokarst lakes from growing. To the south of this border, thick ice thawed as early as the Holocene optimum and lakes existing there are connate. Therefore these lakes gradually become plant-filled. Revegetation ability is high under present-day warming conditions, especially if moss fosters active formation of khasyrei and new freezing of soil. A similar process is observed in the northern part of European Russia, such as in Komi where active growth of moss vegetation results in an increase of erratic water migration to freezing fronts and growth of brae cryogenic terrain.

How Long Can the Present-day Warming Last?

The answer depends on solving the key problem: If the nature of present-day warming results from industrial emissions, we can expect a directional increase of atmospheric and soil temperatures to continue. If it is just a branch of a natural 30-year temperature variation, then atmospheric and frozen ground temperature increase shall stop soon. An in-depth study (Shpolianskaya, 2008) points to the leading role of the natural fluctuation process. One of the last studies (Kononova, 2009) revealed the linkage of atmospheric temperatures in the northern hemisphere from the end of the 19th century to 2008, with the duration of three major types of atmospheric circulation – zonal, north longitudinal and south longitudinal (Figure 14.5). The increased duration of

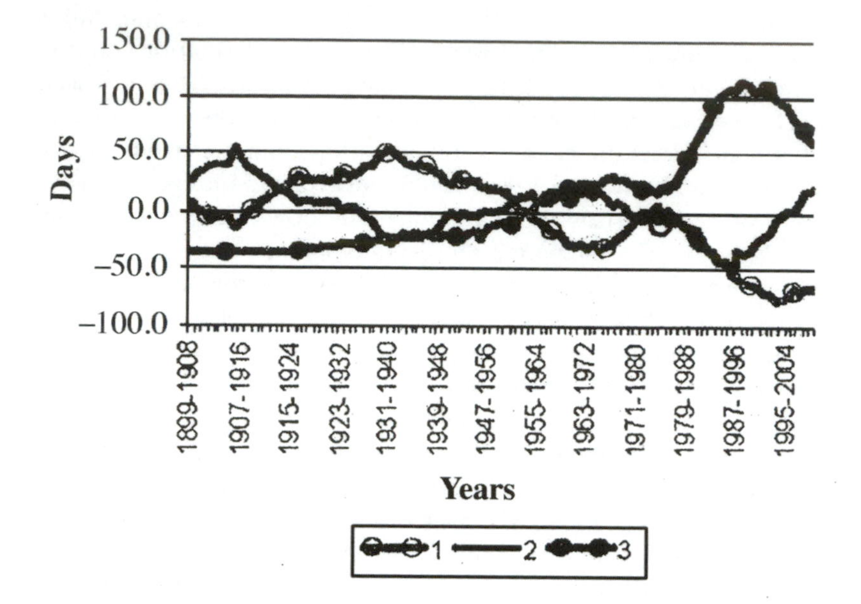

Figure 14.5 *Variations of decennial moving average values of cumulative annual duration of circulation groups from average values for 1899–2008*

Note: 1 – zonal circulation; 2 – north longitudinal circulation; 3 – south longitudinal circulation
Source: Kononova (2009)

zonal processes in 1924–1940s was accompanied by a temperature hike. An increase in north longitudinal processes in 1960–1970s led to cooling on the Earth in general. The sudden warming, which started in the early 1980s, was until now caused by an increased duration of south longitudinal processes and a decreased duration of zonal and north longitudinal processes. In the past year, an apparent inflection of south longitudinal process curve toward decrease has been observed; and a sudden inflection of the north longitudinal process curve toward an increase in duration. This means that warming slowed down and changes of climatic periods are happening. Atmospheric temperature change for Novaya Zemlaya was simulated to assess permafrost behaviour during long-term human-caused warming (Shpolianskaya, 2001, 2008). In accordance with existing forecasts, the simulation was carried out for four warming scenarios – by 1, 2, 3 and 4°C as compared to the present-day atmospheric temperature. The results (Figure 14.6) demonstrate that establishing a new stationary temperature (T) and changing of efficiency from initial (Z_{HAY}) till final (Z_{KOH}), require at least 20,000 years of ongoing climate change.

Thus, even human-caused climate warming most probably will not have serious consequences for cryolithozone. Using the forecasted geothermal classification of western Siberia (Fotiev, 2000) we can assume that, for regions with formation temperatures below –3°C to –4°C, permafrost will not thaw from

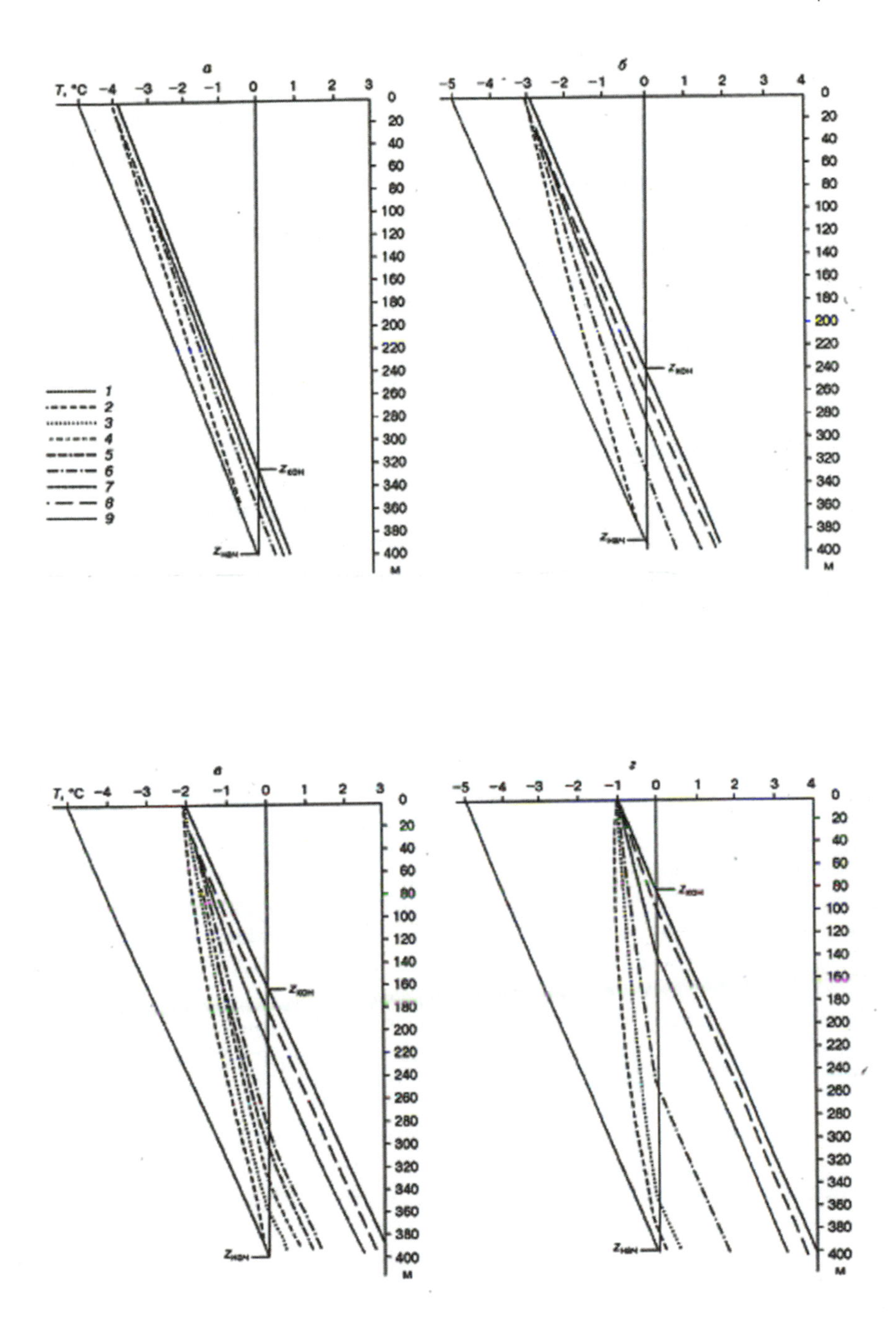

Figure 14.6 *Temporal variations of New Earth permafrost formations temperature and thickness under warming by 1, 2, 3 and 4°C*

Note: 1 – the present-day temperature of frozen soil 400 thick. 2–9 – temperature pattern of geologic formations in: 2 – 1000 years; 3 – 2000 years; 4 – 3000 years; 5 – 4000 years; 6 – 5000 years; 7 – 10,000 years; 8 – 15,000 years; 9 – 20,000 years

the top due to preserved negative temperatures; from below it will thaw in 20,000–30,000 years. In regions located to the south, with permafrost temperatures from 0°C to –3°C, long-term warming can lead to an increase in the seasonal thawing layer and, at the southern border of the cryolithozone, to partial thawing from above. At the same time, according to existing observations (Pavlov, 2002, 2008) and calculations (Fotiev, 2000), it follows that, the higher the reference temperature of the frozen soil, the smaller the affects of climate warming are. From this important finding follows that noticeable reduction of the cryolithozone area due to human-caused warming will take place in its southernmost regions considering the fact that the present-day climate is on the downward branch of a long-period cycle (Figure 14.3).

Overall Conclusion

Due to the ice presence and related phase processes, permafrost as a natural system is generally insensitive to alteration and therefore is quite resistant to climate changes. For a significant alteration of the cryolithozone to occur, directional climate changes over thousands of years are required.

References

Fotiev S. M. (2000) 'Potential changes of geothermal field of cryogenic region of Russia under global climate warming', *Cryosphere of the Earth*, vol IV, no 3, pp14–29

Irpotin S. N., Polischuk Y. M. and Bryksina N. A. (2008) 'Thermokarst lakes area dynamics in continuous and disjunctive cryolithozones of West Siberia under global warming conditions', *Tomsk University Reporter*, no 311, pp185–190

Kirpotin, S., Polishchuk, Yu., Zakharova, E., Shirokova, L., Poprovsky, O., Kolmakova, M. and Dupre, B. (2008) 'One of possible mechanisms of thermokarst lakes drainage in West-Siberian North', *International Journal of Environmental Studies*, vol. 65, no 5, October, pp631–635

Kononova N. K. (2009) 'Classification of circular mechanisms of the Northern hemisphere', in B. L. Dzerdzievsky and M. Voentechizdat (eds) *Geography Environment Sustainability*, pp2–20, http://atmospheric-circulation.ru/wp-content/uploads/2011/01/fluctuations.pdf

Pavlov A. V. (2002) 'Secular atmospheric temperature abnormity in the north of Russia', *Cryoshpere of the Earth*, vol VI, no 2, pp75–81

Pavlov A. V. (2008) *Cryolithozone monitoring*, Academician Publishing House, Novosibirsk

Shpolianskaya N. A. (2001) 'Climatic rhythms and cryolithozone dynamics (evolution analysis in the past and future changes forecast)', *Cryosphere of the Earth*, vol V, no 1, pp3–14

Shpolianskaya N. A. (2008) 'Global climate changes and cryolithozone evolution', Geography college of the Moscow State University, Moscow

Part IV

Governing Biodiversity

15

Governance for Biodiversity: National Plans – the UK Experience

Peter Bridgewater with chapeau by Caroline Spelman MP

Chapeau

I am delighted to welcome this timely publication, as an important contribution to the UN International Year of Biodiversity– a year that saw the very important Tenth Meeting of the Conference of Parties (COP) to the Convention on Biological Diversity (CBD).

The meeting in Nagoya represented both a critical moment in the history of the CBD and in our own human history. We need to consider biodiversity not just for its own sake, but for its mitigation and adaptation roles in the fight against climate change.

We also need to recognize the intrinsic value of biodiversity in the provision of the ecosystem services on which we all rely. Only by recognizing the interdependency of biodiversity, climate change and economic and social development can we hope, as a planet, to overcome the unique challenges we face today.

The chapter that follows explains the UK's experience of biodiversity governance over the past 16 years. I hope it is a useful demonstration of how our understanding of these interdependencies has helped us adapt the ways in which we manage our biodiversity.

As the Secretary of State for the Environment, Food and Rural Affairs in the UK, I am committed to working towards safeguarding our planet's biodiversity for future generations. I look forward to the UK playing an important role in implementing the agreements from the Nagoya meeting.

Introduction

In 1994, the UK became the first country to produce a national biodiversity action plan, following the CBD signed in Rio de Janeiro in 1992.

The UK is interesting from a political perspective, as, although not a federal country, there is considerable devolution of responsibility for biodiversity issues to administrations and agencies in England, Scotland, Wales and Northern Ireland. The UK also provides biodiversity advice and support to its independently governed overseas territories and crown dependencies. A number of Overseas Territories have also developed National Biodiversity Action Plans. In all its approaches, the UK is very diligent in ensuring attention is given to both terrestrial (including freshwater) and marine ecosystems.

Figure 15.1 *Map of the UK*

The pattern of devolution is not equal, therefore there are differences in the way the issues are treated in each of the devolved administrations. In addition, local authorities have heavily devolved responsibilities for land and water management, which gives rise to the need for their stewardship of biodiversity at a local level. And, finally, the UK has a long tradition of individual involvement from amateur naturalists and locally based non-governmental organizations (NGOs).

The UK experience is that effective delivery of national strategies and action plans requires a strong partnership of statutory, voluntary, academic and business sectors, nationally and locally – and this has clear implications for governance.

Much work is delivered through Local Biodiversity Action Plans. This mechanism draws together partners such as local authorities, statutory agencies and NGOs who work together to enhance the quality and distinctiveness of local environments as well as contributing to achieving national targets.

It was recognized at an early stage that biodiversity conservation and sustainable use went beyond simply developing protected areas and species action plans. The UK Biodiversity Action Plan was given a new strategic framework in 2007, entitled *Conserving Biodiversity: The UK Approach* (Defra, 2007), to reflect the changing context brought about by the internationally agreed targets to significantly reduce the rate of biodiversity loss globally by 2010. The value of the ecosystem approach was outlined in this framework as was the overall and increasing impact of climate and other global changes, and the governance effects from increasing devolution within the UK were also key factors.

The integrating framework of the ecosystem approach has 12 principles, of which six are of especial importance in the UK context:

- The objectives of management of land, water and living resources are a matter of societal choices;
- Management should be decentralized to the lowest appropriate level;
- The ecosystem approach should consider all forms of relevant information, including scientific and indigenous and local knowledge, innovations and practices;
- The ecosystem approach should involve all relevant sectors of society and scientific disciplines;
- Recognizing potential gains from management: there is usually a need to understand and manage the ecosystem in an economic context;
- The ecosystem approach should be undertaken at the appropriate spatial and temporal scales.

Taking these six principles and moulding them to a UK context for action on biodiversity they become:

1. developing and interpreting the evidence base;
2. protecting the best sites for wildlife (biodiversity);
3. where cost-effective, targeting action on threatened species and habitats;
4. mainstreaming biodiversity and ecosystem services in all relevant sectors of policy and decision-making;
5. engaging people, and encouraging behavioural change;
6. playing a proactive role internationally.

The governance aspects of each of these are examined in turn, drawing out further lessons from the UK experience.

Developing and Interpreting the Evidence Base

Successive UK governments have recognized that governance for biodiversity needs improved linkages between science and policy, through science–policy interfaces. Science–policy interfaces of different types, sizes and purposes have been, and continue to be, critical forces in shaping the development of governance for biodiversity and ecosystem services. Improving the science–policy interface is a key priority for UK actions in promoting better biodiversity governance. While there are ongoing discussions about the creation of an Intergovernmental Science-Policy Platform on Biodiversity and Ecosystem Services (IPBES) at the international level, within the UK, a science–policy interface for biodiversity already exists in the form of the Joint Nature Conservation Committee (JNCC).

JNCC[1] was created through the 1991 Environmental Protection Act in parallel with the replacement of the UK-wide Nature Conservancy Council with separate agencies in England, Scotland and Wales (accountable to their respective Secretaries of State) to improve the local delivery of nature conservation. JNCC was reconstituted by the Natural Environment and Rural Communities Act 2006.[2] JNCC's role supports devolution, both terrestrially and in the marine environment, providing coherence for European and international policy and devolved implementation.

JNCC brings together members from the nature conservation bodies for England, Scotland, Wales and Northern Ireland and independent members appointed by the Secretary of State for the Environment, Food and Rural Affairs under an independent chair. Support for the committee's work is provided by around 130 staff with scientific and technical expertise, extensive evidence-based knowledge at global, European and national levels and skills in working with other organizations. In recent years, JNCC has become very involved in work on marine biodiversity, working with devolved administrations and the newly created Marine Management Organisation.

A sound evidence base is essential to support effective biodiversity conservation. Assessment of the existing knowledge base, and commissioning new and innovative research and associated monitoring is essential for:

- assessing the current status and trends in biodiversity;
- understanding the value of biodiversity and ecosystem services;
- understanding the reasons for unfavourable status and decline in biodiversity;
- assessing future vulnerability and identifying effective remedial measures and strategies;
- assessing the outcomes and effectiveness of policy; and
- innovation in the way we collect, manage and use evidence to support policy and action.

The National Biodiversity Network – a charitable trust supported by public, private and volunteer members, has been established to provide a single point of access to the more diffuse but nevertheless very important sources of this evidence and the technology developed has provided the foundation for the Global Biodiversity Information Facility.

The UK Biodiversity Research Advisory Group (UK BRAG) was set up in 2003 by the UK Biodiversity Partnership to:

- identify, promote and facilitate biodiversity research to support biodiversity action plan commitments;
- coordinate effective and efficient UK engagement with European biodiversity research issues;
- contribute to effective biodiversity research networking in the UK; and
- support knowledge transfer activities in relation to biodiversity research.

UK BRAG provides a forum for developing strategic research, ensuring there is coordination between the main UK funders of biodiversity evidence and providing an opportunity for the science community to contribute to the development of policy-relevant research programmes.

The UK has developed a set of 18 indicators, with 33 component measures, to summarize some of the key priorities for biodiversity in the UK. The UK Biodiversity Indicators were first published in the National Statistics publication *Biodiversity Indicators in Your Pocket 2007* and subsequently updated in May 2008, April 2009 and May 2010 (Defra, 2010). The study containing the indicators is supported by a website (www.jncc.gov.uk/biyp), which provides more detail about these high-level indicators and provides access to component data if required.

As elsewhere, the UK has found indicators are a valuable way of summarizing and communicating evidence, maintaining momentum on implementing policy initiatives, and thus providing better biodiversity governance at many levels.

Protecting the Best Sites for Biodiversity (Wildlife)

Protected areas are an important element of the UK approach to conservation, delivered through identifying the most important areas, providing legal protection, influencing activities within their boundaries and repairing damage that occurs within them. These sites are core to our strategy to ensure that biodiversity is able to mitigate against, and adapt to, environmental changes, particularly climate change.

Around 10 per cent of the land area of the UK is notified as Sites of Special Scientific Interest (SSSIs) or Areas of Special Scientific Interest (ASSIs). Many are also Special Areas of Conservation (SACs) under the European Commission (EC) Habitats Directive, Special Protection Areas (SPAs) under the EC Birds Directive or Wetlands of International Importance under the Ramsar

Focal area, indicator, title and individual measures (where applicable)		Long-term change[1]	Change since 2000
Focal area 1. Status and trends of the components of biological diversity			
1a. Pop. of selected selected species (birds)	Breeding farmland birds	✗ 1970–2008	✗
	Breeding woodland birds	✗ 1970–2008	✓
	Breeding water and wetland birds	≈ 1975–2008	≈
	Breeding seabirds	✓ 1970–2008	✗
	Wintering waterbirds	✓ 1975–2007–8	✗
1b. Pop. of selected species (butterflies)	Semi-natural habitat specialists	✗ 1976–2009	≈
	Generalist butterflies	≈ 1976–2009	≈
1c. Populations of selected species (bats)		✗ 1978–1992	✓
2. Plant diversity selected species (birds)	Arable and horticultural land	✓ 1990–2007	✓
	Woodland and grassland	✗ 1990–2007	✗
	Boundary habitats	✗ 1990–2007	✗
3. UK priority species		•••	✓
4. UK priority habitats		•••	≈
5. Generic diversity	Native sheep breeds	•••	≈
	Native cattle breeds	•••	✓
6. Protected areas	Total extent of protected areas	✓ 1996–2009	✓
	Condition of A/SSSIs	•••	✓
Focal area 2. Sustainable use			
7. Woodland management		•••	✓
8. Agri-environment land	Higher level, targeted schemes	✓ 1992–2009	✓
	Entry type schemes	•••	✓
9. Sustainable fisheries		✓ 1990–2008	✓
10. Impact of air pollution	Acidity	✓ 1996–2005	≈
	Nitrogen	✓ 1996–2005	≈
11. Invasive species	Freshwater species	✗ 1960–2008	≈
	Marine species	✗ 1960–2008	✗
	Terrestrial species	✗ 1960–2008	✗
12. Spring index		Not assessed	Not assessed

Figure 15.2 *UK biodiversity indicators*

Source: Defra (2010)

Focal area, indicator, title and individual measures (where applicable)		Long-term change[1]	Change since 2000
Focal area 4. Threats to biodiversity			
13. Marine ecosystem integrity		deteriorating 1968–2008	little or no overall change
14. Habitat connectivity	Broad-leaved, mixed and yew woodland	insufficient or no comparable data	insufficient or no comparable data
	Neutral grassland	insufficient or no comparable data	insufficient or no comparable data
15. Biological river quality		improving 1990–2008	improving
Focal area 5. Status of resource transfers and use			
16. UK biodiversity expenditure		insufficient or no comparable data	improving
17. UK global biodiversity expenditure		insufficient or no comparable data	improving
Focal area 6. Public awareness and participation			
18. Conservation volunteering		insufficient or no comparable data	improving

Key: improving; little or no overall change; deteriorating; insufficient or no comparable data

1 The earliest available year is used as the baseline for assessment of long-term change. The base year used for each measure is shown in the table. Where data are unavailable, or do not precede 1996, a long-term assessment is not given.

Convention on Wetlands. The overall total extent of land and sea protected in the UK has increased from 2.3 million to 3.8 million hectares between 1996 and 2009 – an increase of 62 per cent (Figure 15.1).

Since 2005, the percentage of sites in favourable or recovering condition in the UK has increased by 15 per cent for ASSI/SSSIs, 11 per cent for SACs and 9 per cent for SPAs (Figure 15.2). Large increases are noted in England, with 93.1 per cent of SSSIs now in favourable or improving condition compared with 57 per cent in 2003, with more modest changes in Scotland and relatively little change in Northern Ireland.

Table 15.1 *Assessment of change in area and condition of UK protected areas*

	Long term	Since 2000	Latest year
Total extent of protected areas	improving 1996–2009	improving	Increased (2009)
Condition of ASSI/SSSIs	insufficient or no comparable data	improving	Increased (2009)

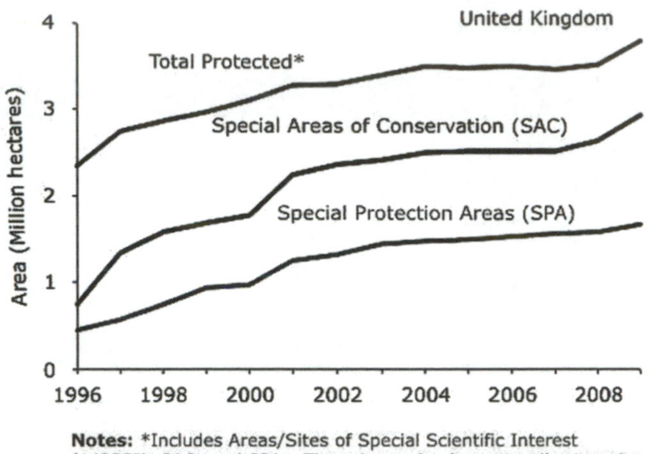

Figure 15.3 *Extent of nationally and internationally important protected areas: i) total extent; ii) Special Areas of Conservation; iii) Special Protection Areas, 1996–2009*

Source: Joint Nature Conservation Committee, Natural England, Countryside Council for Wales, Northern Ireland Environment Agency, Scottish Natural Heritage

Over time, the species and ecosystems present at any individual site may change, but still remain valuable assets for biodiversity conservation. Making sites more robust to environmental change – by improving their quality and condition, reducing the impact of other pressures in the surrounding areas, buffering and where appropriate making them larger – is a priority. Governance arrangements that take into account not just protected areas, but their integration with the wider land and seascape, and promotes such areas as part of a living landscape, are part of the UK approach. In part, this is less complex than for some countries because the protected area network is mostly comprised of very small sites, with the larger areas being International Union for Conservation of Nature (IUCN) protected area category V – working landscapes.

In the UK, governance mechanisms to achieve management for biodiversity change come from actions at the level of the conservation agencies for England, Northern Ireland, Scotland and Wales, working through the JNCC when issues have a UK-wide or international aspect, such as reporting to the European Union (EU), CBD and other relevant regional and international agreements dealing with terrestrial and marine biodiversity.

Targeting Action on Threatened Species and Habitats

Not all of the UK's priority species and habitats are found in designated areas, and experience shows that targeted action can deliver sustained improvements

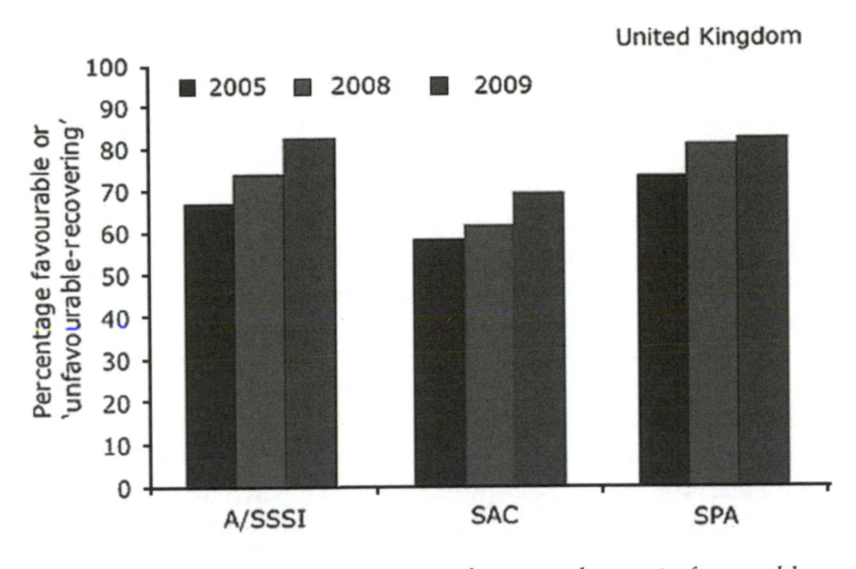

Figure 15.4 *Cumulative proportion of protected areas in favourable or 'unfavourable recovering' condition, by feature or by area, 2005–2009*

Note: 1) No data available for Wales; 2) England figures based on area: Scotland, Wales and Northern Ireland figures based on number of features.
Source: Joint Nature Conservation Committee, Natural England, Countryside Council for Wales, Northern Ireland Environment Agency, Scottish Natural Heritage

to the status of species and habitats. EU and domestic legislation provides a solid base on which to build this action. It provides statutory protection for certain species and habitats, and requires EU member states to take special conservation measures for particular species and habitats in order to maintain them at or restore them to 'favourable Conservation Status'.

To support and supplement this, the UK Biodiversity Partnership has identified a UK list of priority species and habitats. The UK list of priority species and habitats,[3] which was updated in 2007, is an important reference source. The four devolved administrations in the UK work together to translate these priorities into programmes of work delivered by partnerships of statutory, voluntary, academic and business organizations at the level most appropriate to the needs of biodiversity, again with the JNCC playing a key role as science–policy interface.

Since the publication of the UK Biodiversity Action Plan in 1994, one of the features of the UK approach has been the agreement and publication of specific targets for certain species and habitats.[4]

Usually the action to achieve a species target drives habitat action that is beneficial for a much wider range of species. For example, conservation efforts to restore populations of the bittern (*Botaurus stellaris*), a UK priority species that was formerly thought to be extinct, have resulted not only in increasing its numbers, but also in the creation of more than 800ha of reedbeds. Similarly, conservation action for the large blue butterfly (*Maculinea arion*), delivered through an agri-environment scheme has resulted not only in increasing the population of this butterfly – formerly thought to be extinct – to 40,000 adults, but also in bringing grassland sites back into appropriate grazing management. Overall, it makes sense to manage habitats for the range of niches that species need, in a landscape-scale approach, managed through approaches at devolved administration level, with UK context provided by the JNCC.

The Millennium Ecosystem Assessment (MEA, 2005) highlighted the relationship between ecosystems and human well-being and the need to take action to reverse ecosystem degradation by addressing the key drivers and valuing ecosystem services. The UK has tried to reconcile lists of priority species with the wider agenda of maintaining ecosystem services.

One response is the National Ecosystem Assessment. This is a first analysis of the UK's natural environment in terms of the benefits it provides to society and continuing economic prosperity. Its primary aims are: (a) to provide a high level picture of the current state and trends since 1945 of the UK's ecosystems and ecosystem services; and (b) to look to the future (2060) to evaluate change under plausible scenarios and consider a range of policy options. It is an inclusive process involving many governmental, academic, NGO and private sector institutions. Preliminary findings were published on the website[5] in February 2010. The final products are scheduled for launch in February 2011.

Following devolution from 1998 onwards, targets that had previously been expressed at UK level were agreed at individual administration level. Each devolved administration has been engaged in identifying how and where they can contribute to meeting national targets, and in setting targets for their own areas. In order to take forward the agreed approach of achieving biodiversity enhancements at a landscape scale, national to local levels of the partnership are working together to identify landscape-scale projects in each of the countries of the UK. A key lesson for governance arrangements for biodiversity in the UK is that devolution requires engagement between national, regional and local levels of decision-making.

An example: Linking national and local biodiversity planning in Scotland

There are two strands to planning biodiversity work in Scotland – one operating at the country level and one at the local level.

At the country level, strategic planning for biodiversity is carried out within five Ecosystem Plans covering each of the main ecosystem types in Scotland. These plans address the key issues relating to the ecosystems, their constituent habitats and species, and the services they provide, in an integrated way. The focus is on developing broad landscape-scale actions that improve the health of ecosystems.

At the local level, planning for biodiversity action on the ground is coordinated by Local Biodiversity Action Partnerships (LBAPs). LBAP plans are the principle mechanism for expressing biodiversity priorities at a local level in

Scotland. Each plan covers one or more local government areas and most contain a mixture of national priorities and locally important habitats and species.

To make these two processes more coherent, Scottish Natural Heritage, in partnership with the Macaulay Land Use Research Institute and the Royal Botanic Gardens, Edinburgh, are developing an ecologically based framework to help LBAP partners prioritize actions for biodiversity in their area. Results from this project will help to communicate national priorities at the local level, develop better governance at all scales and help to focus resources more effectively. And, of course, the Scottish work then supports the UK focus on European and international activities.

Mainstreaming Biodiversity and Ecosystem Services in All Relevant Sectors of Policy and Decision-making

The way people use terrestrial and marine biodiversity has led to many of the acknowledged declines in UK biodiversity. Improving governance for biodiversity is helping to reverse many of these declines, enhance the conditions for a wide range of other wildlife, and sustain ecosystem services that will ultimately reduce costs to other sectors.

Work to mainstream consideration of biodiversity and ecosystem services is achieved through the biodiversity or environment strategies of each of the four devolved countries within the UK (England, Northern Ireland, Scotland and Wales) and through their statutory country conservation bodies as the main delivery agents.[6] At both UK and devolved administration level, implementing the strategies is a cross-government responsibility. To reduce the rate of biodiversity loss, strategies seek to make biodiversity part of the mainstream in policy development and emphasize that healthy, thriving and diverse ecosystems are essential to everybody's quality of life and well-being.

Public bodies have an important role in contributing to biodiversity, and domestic legislation nationally and regionally in the UK now includes a biodiversity duty on public bodies.[7]

Engaging People and Encouraging Behaviour Change

Reducing biodiversity loss has widespread public support and a great deal of biodiversity conservation is achieved by enthusiastic volunteers. But we need to engage more people in taking action to maintain and enhance biodiversity as part of their everyday lives. In doing so, we assist in delivering the objectives of the CBD, which include incorporating the need for conservation into communication, education and public awareness programmes. Between 2000 and 2009 there was a 51 per cent increase in time spent volunteering, and in 2009 the total time spent was equivalent to around 1 million working days, based on an index of the number of hours worked by volunteers in eight major UK conservation charities and the conservation agency for England, Natural England (Figure 15.5).

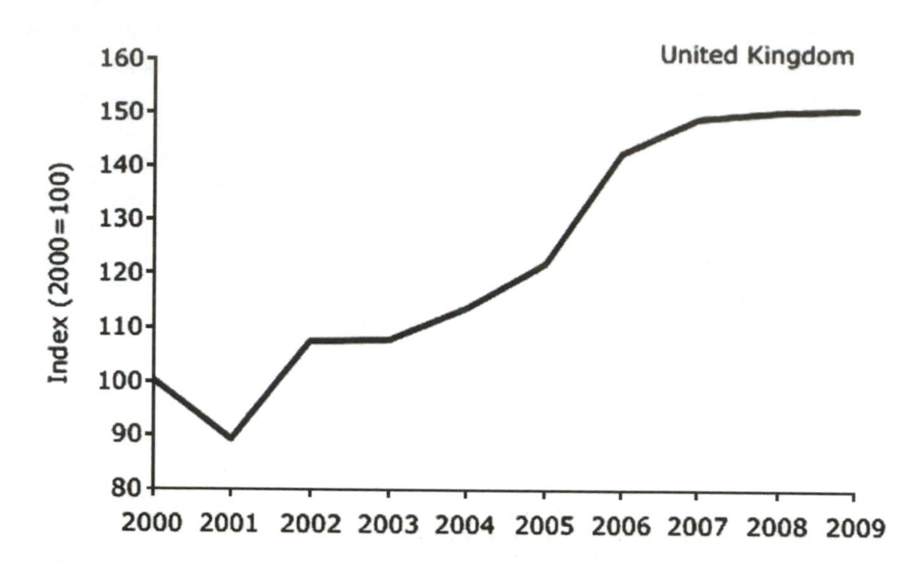

Figure 15.5 *Index of volunteer time spent in biodiversity conservation in selected UK conservation organizations, 2000–2009*

Note: 1) Interpolated data have been used by Defra to fill missing years for Woodland Trust (2000–2001), Butterfly Conservation (2000–2002) and The Wildlife Trusts (2000 to 2004); 2) As data provided by the Royal Society for the Protection of Birds (RSPB) were for financial years as opposed to calendar years, 2008–2009 data was allocated to 2008 and Defra estimates were made for 2009.
Source: Bat Conservation Trust, British Trust for Conversation Volunteers, British Trust for Ornithology, Butterfly Conservation, Natural England, Plantlife, Royal Society for the protection of Birds, The Wildlife Trusts, Woodland Trust

Table 15.2 *Assessment of change in volunteer time spent in biodiversity conservation*

	Long-term	Since 2000	Latest year
Conservation volunteering	●●●	✓	Increase (2009)

See key to symbols after figure 15.2

Playing a Proactive Role Internationally

Internationally, the UK is actively involved in multilateral environmental agreements, at global and European scales. The principal mechanism by which the UK government assists the aims of the CBD globally is the Darwin Initiative. Announced by the UK government at the Rio Earth Summit in 1992 and launched in 1993, it assists countries that are rich in biodiversity but poor in financial resources. While the CBD is a major focus, meritous collaborative projects that draw on UK biodiversity expertise and that assist implementation under the Convention on International Trade in Endangered Species (CITES) and the Convention on Migratory Species (CMS) are also supported where funding is available.

Thus far, the Darwin Initiative has committed more than £79 million to 725 projects in 156 countries across the developing world. Sustainability is a key issue for consideration on many projects and the type of project funded has a significant impact on the level and type of sustainability that can be achieved.

Conclusion

Over the past 16 years of the UK Biodiversity Action Plan, we have adapted our strategy to respond to devolution within the UK, our growing awareness of the challenges posed by climate change, and the need to work within the ecosystem approach. We have learnt many lessons along the way. We have achieved many successes, but there is still much more to do. As we redouble our efforts to meet a new suite of challenging targets for 2020, we will place renewed emphasis on linking biodiversity and the provision of ecosystem services, understanding the value of biodiversity, and seeking to make biodiversity part of all sectoral policies.

Acknowledgements

With thanks for input from Ant Maddock, Tara Pelembe, Paul Rose, Sarah Webster and James Williams.

Notes

1 www.jncc.gov.uk/page-5288
2 www.opsi.gov.uk/acts/acts2006/ukpga_20060016_en_1
3 www.ukbap.org.uk
4 www.ukbap.org.uk/GenPageText.aspx?id=98
5 http://uknea.unep-wcmc.org/
6 Natural England, Scottish Natural Heritage, the Countryside Council for Wales, and the Environment and Heritage Service
7 S1 of Nature Conservation (Scotland) Act 2004, www.opsi.gov.uk/legislation/scotland/acts2004/20040006.htm; Section 40 of the Natural Environment and Rural Communities Act in England and Wales, www.opsi.gov.uk/ACTS/acts2006/20060016.htm; and Wildlife (Amendment (Northern Ireland) Order), www.opsi.gov.uk/si/si1995/Uksi_19950761_en_1.htm

References

Defra (Deparment for Environment, Food and Rural Affairs) (2007) *Conserving Biodiversity: The UK Approach*, Department for Environment, Food and Rural Affairs, London, www.defra.gov.uk/environment/biodiversity/documents/conbiouk-102007.pdf

Defra (2010) *UK Biodiversity Indicators in Your Pocket 2010*, Department for Environment, Food and Rural Affairs, London, www.jncc.gov.uk/pdf/BIYP_2010.pdf

MEA (Millennium Ecosystem Assessment) (2005) *Ecosystems and Human Well-being: Biodiversity Synthesis*, Island Press, Washington, DC

16

Governing Biodiversity

Felix Dodds and Richard Sherman

> *We need a United Nations-wide global biodiversity strategy that ensures a swift passage from the destruction of nature to the prevention of the further degradation of the natural resources and ecosystem services that underpin all life on Earth. This is an investment for our children. And this will support the United Nations' efforts to fight poverty.* (Jochen Flasbarth, Chair of the Working Group on Review of Implementation of the Convention on Biological Diversity (CBD))

The model of economic growth that has been prevalent since 1972 and accelerated since globalization took root after 1992 should be called the period of irresponsible capitalism. The benefits from the actions have been for the few, increasing the gap between the wealthiest and poorest in society. It has been a period of privatizing the use of natural resources and socializing the impacts with little or no regard for the sustainability of the natural resource base upon which we live and on which our wealth is built.

In many areas the destruction of our biodiversity is increasing. In *Nature* magazine (Rockström et al, 2009) a group of scientists looked at nine areas they felt were needed to recognize the limits. In addition to each area being important in its own right, they were also deemed interconnected; so what happens when one impacts the others? (Rockström et al, 2009). The nine areas that were studied include: climate change, stratospheric ozone, land-use change, freshwater use, biological diversity, ocean acidification, nitrogen and phosphorus inputs into the biosphere and oceans, aerosol loading and chemical pollution. It was found that biodiversity loss should be no more than ten times the background rates of extinction; the current species loss stands between 100 and 1000 times the natural rate. This is perhaps a greater threat to humanity than climate change. As biodiversity loss increases, ecosystems will fold and our food chains, and possibly our health, will be negatively affected. According to the United Nations Environment Programme's (UNEP) chief scientist Professor Joseph Alcamo, there remain perhaps even greater impacts as we uncover the interlinkages between ecosystems we had not previously realised were connected.

'Since 1970, we have reduced animal populations by 30%, the area of mangroves and sea grasses by 20% and the coverage of living corals by 40%' (Alcamo, 2010). Clearly science is telling us that we have a serious problem. But is this another example of politicians being unable to adequately address the challenges we face? Even when confronted by clear and decisive science? A key component of the era of irresponsible capitalism has been the consequential accountability and governance crises.

In 2001 and 2002, the international community adopted as a benchmark the target of significantly reducing biodiversity loss by 2010. However, as the international community marked the International Year of Biodiversity, not one government had met this target and many were far off track. Underscoring this, the 2010 Global Biodiversity Outlook (GBO3) warned that it is likely we will see 'a massive further loss of biodiversity', and that this will result in 'severe reduction of many essential services to human societies as several "tipping points" are approaching'. Furthermore the report said that 'ecosystems shift to alternative, less productive states from which it may be difficult or impossible to recover' (GBO3, 2010). The convention has played an important role in keeping the conversation on biodiversity in the minds of governments, stakeholders and the media. But the convention has not been able to play the role it should have because it does not wield regulatory powers. The convention does not require mandatory national reporting; if a government does not report there is no sanction. This is just one example of the institutional crisis facing the international community. A crisis further compounded by the fact that the institutions set up in the last 40 years such as UNEP and in particular the Commission on Sustainable Development (CSD) were never given the mandate to enforce agreements and hold member states to account.

The success of the Rio Summit in agreeing two conventions relating to biodiversity and climate and setting in motion the process for a third on desertification should not be underestimated. Despite this, Rio highlighted the clear problem of fragmentation within the system, an issue that the discussions and negotiations on international environmental governance have been trying to address since then. There are now more than 700 environmental conventions, charters, agreements, accords, protocols and treaties in force, from global to regional to bilaterally applicable agreements (UNEP, 2007). The fragmentation is not just at the global level but is often reflected at the governmental level as well. What has been very clear since the Rio Earth Summit is that governments are not very good at joined-up thinking, or acting for that matter. The lack of delivery of the biodiversity target and most of the Millennium Development Goals (MDGs) are clear examples of that.

Biodiversity Governance Post-2010

With the year 2010 bringing a number of important opportunities for elevating the biodiversity discussion, coupled with the 2012 United Nations (UN) Conference on Sustainable Development, the so-called Rio+20 meeting, this section looks at a number of suggestions that we believe would ensure a more coherent, less fragmented and more productive approach to biodiversity within the UN system. The challenges we will face due to increasing biodiversity and ecosystem insecurity over the coming decade perhaps make this, along with climate change and poverty, one of the major challenges that political leaders need to urgently address. In this section, we look at several options for strengthening the sustainable development architecture, based on an elevated role for biodiversity. We do not present options regarding the CBD itself, rather we take a broader look at the institutional framework for sustainable development with the aim of identifying areas where the biodiversity community can be active players in advancing a more coherent international governance architecture.

A UN sustainable authority

Since its adoption in 1993, the UN General Assembly's Second Committee has adopted annual resolutions on the CBD, based on reports submitted by the Executive Secretary. Through the Millennium Summit in 2000, the World Summit on Sustainable Development (WSSD) in 2002 and the 2005 World Summit Outcome Document the UN General Assembly has placed biodiversity high on its agenda. Following the WSSD, the 2010 biodiversity target was added to MDG 7 on environmental sustainability. In 2006, the UN General Assembly declared 2010 as the International Year of Biodiversity under the theme 'biodiversity for development and poverty alleviation' and, in 2008, it further elevated the status of biodiversity and its link to poverty eradication by agreeing to host a high-level meeting of the UN General Assembly with the participation of heads of state prior to the 2010 review of the MDGs. Some have seen this as the last opportunity for governments to get back on track to meet the 2015 targets.

In 2002, the UN General Assembly adopted sustainable development as a key element of the overarching framework for the UN, particularly for achieving the internationally agreed development goals, including those contained in the Millennium Declaration. The Johannesburg Plan of Implementation (JPOI), under the chapter on the institutional framework for sustainable development, called on the UN General Assembly to give overall political direction to the implementation of Agenda 21 and its review. The World Summit 2005 Outcome Document also committed member states to promote the integration of the three components of sustainable development – economic development, social development and environmental protection – as interdependent and mutually reinforcing pillars. The UN General Assembly further reaffirmed that poverty eradication, changing unsustainable patterns of production and consumption and protecting and managing the natural resource base of economic and social development are overarching objectives of, and essential requirements for, sustainable development.

The CSD has, over the years, proved to be an important 'home' for keeping the broad sustainable development agenda under active review, and has been instrumental in launching a number of new initiatives and securing intergovernmental cooperation. The CSD is the highest-level forum for sustainable development in the UN system and as such has addressed biodiversity and the convention in the context of sustainable development four times, in 1995, 1997, 2000 and 2002. At key points, the CSD has played an important role in giving impetus to the work around biodiversity. In 1995, at the UN General Assembly Special Session (UNGASS), it enabled the first review of progress since the convention came into force in 1994. In 2000, the CSD was again able to make comments on the coming into force of the first protocol for the Convention on the Cartagena Protocol on Biosafety. Perhaps the most important was the ten-year review in 2002, with the adoption of the 2010 biodiversity target along with a number of other significant measures (UN, 2002).

While the intergovernmental mandate on sustainable development is clear and has been unanimously agreed to, the development and maturation of the UN's pillar organizations have stagnated. Despite the successes of the CSD, it remains an institutionally and politically weak global authority on sustainable development. The CSD has been unable to deliver on its original mandate (of monitoring and reviewing Agenda 21 implementation) and on its assumed role (of negotiating decisions that move sustainable development forward) (Najam et al, 2006). The deviation from the original mandate, particularly as it relates to financing the means to implement an overall direction on sustainable development to the UN system, remains a concern for many participants.

Despite being the highest-level forum on sustainable development, the CSD has struggled to appropriately address the three pillars of sustainable development, and as a functional commission of the UN Economic and Social Council (ECOSOC) often lacks the appropriate political decision-making powers of the UN General Assembly or its subsidiary organs. The possibilities for the CSD to successfully carry out its mandate are in many ways shaped and defined by its traditional, bureaucratic and institutional surroundings. Its power and limitations derive in large part from its place within the UN system (International NGO Task Group on Legal and Institutional Matters (INTGLIM, 1997)). The CSD and its parent body, ECOSOC, are currently failing to effectively oversee the system-wide coordination of sustainable development, as well as the balanced integration of economic, social and environmental aspects of the UN's policies and programmes. Attempts to reform both bodies have provided limited remedial benefits; the renewed focus on system-wide coherence in light of the World Summit 2005 and the urgent need to meet the MDG and JPOI targets point to a problem with the current system. The lack of an operative function, particularly at the country level, for sustainable development within the UN system, and the fragmented coordination of the UN's work has led to overlap, duplication, confused priorities and 'turf wars'. The CSD has also not managed to influence the processes that govern development today, such as economic and trade policies.

The present state of discussions on sustainable development within the CSD and the UN General Assembly suggests that the present set-up is not working. A number of suggestions should be considered which would enable a more coherent approach to sustainable development within the UN, and therefore in relation to biodiversity and ecosystem insecurity and other emerging and critical threats. These include:

- Upgrading the UN Commission on Sustainable Development to a Council of the UN General Assembly;
- Creating a Sustainable Development Security Council; and
- Transforming the UN Trusteeship Council into an Ecological Security Council.

The transformation of the CSD into a Council of the General Assembly is following the lead of human rights, which did this in 2007. The idea of creating a sustainable development Security Council might go hand in hand with German chancellor Angela Merkel's suggestion of creating of an 'economic council' at the UN alongside the Security Council, and a UN charter on sustainable economics. If the chancellor's suggestion does go ahead, such a council could also become the overarching body for coordinating all sustainable development financial mechanisms.

The environment pillar of sustainable development

As the leading environmental authority within the UN, the UNEP has been instrumental in the development of a wide range of international agreements and conventions concerned with advancing protection of the world's biological diversity and the ozone layer, as well as the sound management of chemicals and persistent organic pollutants. However, for more than 38 years, government and independent analysis of UNEP's impact have regularly identified issues that beleaguer the current system. Areas identified as key weaknesses in the existing arrangements on International Environment Governance (IEG), include: inadequate level of integration of environmental considerations into the mainstream of decision-making; inadequate approaches to global environmental management; impacts of globalization; fragmented machinery; institutional mandates and environmental agreements predominantly follow a sectoral approach to environmental management; weak international dispute mechanisms; and lack of a holistic approach to international environmental governance. At the political level, areas most singled out for further attention include: the discrepancy between commitments and action; and the lack of a strong political base has contributed to a failure to effectively mainstream and integrate the environment into the wider macro-economic arena, and particularly within the World Trade Organization (WTO). This has lead to fragmentation, limited financial resources, poor enforcement of multilateral environmental agreements as well as an imbalance between international environmental governance and other international trade and financial regimes.

In order to address these issues, we suggest a reinvigorated process to focus on UNEP's Global Ministerial Environment Forum (GMEF) and strengthening UNEP.

UNEP Global Ministerial Environment Forum (GMEF)

One of the outcomes of the 2000–2002 UNEP-driven IEG process, was the recognition that the 'proliferation of institutional arrangements, meetings and agendas, while having the benefit of specialization, may weaken policy coherence and synergy and put further strain on limited resources'. However, since then not much progress has been made in the area of enhanced coordination among the Multilateral Environmental Agreements. Following the intervention of former Secretary-General Kofi Annan, the UN General Assembly agreed to establish, under the authority of the UNEP Governing Council, the Global Ministerial Environment Forum (GMEF) as the high-level environmental policy forum within the UN system. The GMEF's mandate is clear: 'Provide general policy guidance for the direction and coordination of environmental programmes and make cross-cutting recommendations to other bodies while respecting the independent legal status and autonomous governance structures of such entities.' However, the GMEF's authority has diminished significantly over the past few years. This is partly due to its failure to provide leadership and direction, particularly with regard to system-wide coordination as stipulated in the 2002 IEG report, which suggested that the GMEF 'should identify ways and means of improving and strengthening its interrelationship with autonomous decision-making bodies, such as conferences of the parties to [Multilateral Environmental Agreements] MEAs'. Going forward, the GMEF should become a forum concentrating on dealing with serious threats to the environment and a platform for Ministers of Environment to speak out forcefully on these environmental challenges.

A strengthened UNEP

UNEP has a considerable role in the area of biodiversity by overseeing the conventions, additionally it produces Global Environmental Outlook (GEO) and Global Biodiversity Outlook (GBO). It runs programmes for the Great Apes Survival Project, Global Programme of Action (GPA) for the Protection of the Marine Environment from Land Based Activities, UNEP-GEF Project on Development of National Biosafety Frameworks and it has collaborating centres such as the UNEP World Conservation Monitoring Centre (WCMC). UNEP is also increasingly developing work on biodiversity and ecosystem economics.

Discussions on the complexity of the IEG process have clearly favoured strengthening UNEP's Nairobi headquarters as the lead UN body responsible for all environmental programmes and activities within the UN system. These discussions have also noted that the achievement of progressive decisions on environmental and sustainable development issues sometimes requires more political will than is available to all governments. There is an urgent need for a stronger UNEP as the international authority on the environment to safeguard the environmental pillar of sustainable development. A strengthened UNEP should therefore have a new mandate. Such a mandate should build greater coherence between environmental and social agendas, making the concept of 'environment for development' a reality. It would act as a platform for both standard setting and other interaction with national, international and UN bodies. The principles of cooperation and of common but differentiated responsibilities should be reflected in the implementation of this revised mandate.

The economic pillar of sustainable development

Financing biodiversity protection or developing an approach to ecosystem services is at the heart of trying to reduce the loss of biodiversity and the destruction of ecosystems. It is worth recalling that in JPOI, Chapter 3, paragraph 44 (a) the international community stressed, 'in particular, the need to integrate the objectives of the Convention into the programmes and policies of the economic sectors of countries and international financial institutions'. We are concerned that there needs to be a better link between the Bretton Woods Institutions and the ECOSOC, both in relation to guidance on system-wide activities related to the economic pillar of sustainable development, and in relation to the creation of a more coherent, transparent and accountable system for financing biodiversity protection. This concern relates not only to biodiversity issues, but to the fragmented approach ECOSOC and the Bretton Woods Institutions have towards sustainable development issues in general.

Our suggestions would therefore be:

- A sustainable development segment as part of the annual ECOSOC Substantive Session, with a dedicated annual debate on thematic issues such as biodiversity, climate change and desertification. This general idea was first introduced in the UN Secretary-General's 2007 Report on System-Wide Coherence, but has gained little traction in diplomatic circles. However, the clear failure to make any significant progress on the 2010 targets suggests that a more integrated approach to biodiversity and sustainable development is required, particularly at the ECOSOC level, which is the UN's principle organ dedicated to coordinating economic, social and related work of the 14 UN specialized agencies, functional commissions and regional commissions. Furthermore, ECOSOC receives reports from the 11 UN funds and programmes, providing the appropriate space for further consideration of sustainable development policy coherence. If such a proposal were to be adopted, it would be timely to have the first dedicated session focusing on how to meet the biodiversity target by 2015, as well as making input into the design of the post-2015 MDGs.
- An additional day of the Special High-Level Meeting of ECOSOC with the Bretton Woods Institutions, WTO and UN Conference on Trade and Development (UNCTAD), to specifically deal with sustainable development finance, in particular the Rio Conventions: biodiversity, climate change and desertification. It is important to stress that discussions do not duplicate the

finance discussions under the Conference of the Parties (COPs), which address operational and programmatic issues, but instead look at innovative sources of finance related to climate change, biodiversity and desertification. The recent findings of The Economics of Ecosystems and Biodiversity (TEEB) study has drawn significant attention to the global economic benefits of biodiversity by highlighting the growing costs of biodiversity loss and ecosystem degradation. It draws together expertise from the fields of science, economics and policy to enable practical action moving forward.

An umbrella intergovernmental panel on sustainable development

The Millennium Ecosystem Report and UNEP's GBO played a significant role in highlighting the scientific knowledge on the links between biodiversity, ecosystem services and human well-being. There was a clear need to bring this together and to enable emerging scientific knowledge to be translated into specific policy action at the appropriate levels. The biodiversity and ecosystem related Multilateral Environmental Agreements provide for scientific and technical cooperation. However, it was felt that this could be further enhanced by creating an Intergovernmental Science-Policy Platform on Biodiversity and Ecosystem Services (IPBES) based on the model of the Intergovernmental Panel on Climate Change (IPCC). This would then provide a mechanism 'that could provide a scientifically sound, uniform and consistent framework for tackling changes to biodiversity and ecosystem services' (IPBES website, www.ipbes.net/about-ipbes.html).

The IPCC is seen as a model of providing governments with independent scientific advice and information. Even with 'climategate' and other attempts to discredit the work of the IPCC, it stands as a great example of how science can help policy-makers. It is therefore not surprising that there have been attempts to duplicate in other areas such as the Intergovernmental Panel on Forests (IPF) set up in 1995 attempted to collect: 'Scientific research, forest assessments and the development of criteria and indicators for sustainable forest management' (www.unclef.com/esa/forests/ipf_iff.html). The IPF very quickly became a political rather than a scientific forum. This was recognized and the organization changed its name in 1997 to the Intergovernmental Forum on Forests (IFF). There have also been suggestions by some to set up a high-profile independent panel on desertification, much like the present IPCC. This is also the backdrop to the setting up of the Intergovernmental Science-Policy Platform on Biodiversity and Ecosystem Services (IPBES, which has already been discussed).

In 2005, the then President of France, Jacques Chirac, announced a proposal for a consultative process towards the establishment of an International Mechanism of Scientific Expertise on Biodiversity. Between 2006 and 2007 numerous meetings were held, and at its final session in 2007, participants adopted a statement in which they invited the UNEP executive director to convene an intergovernmental meeting to consider establishing an effective science–policy interface. Following this decision, adopted by UNEP's Governing Council in 2009 (Decision 25/10) and its Special Session in 2010 (Decision

SS.XI/4), progress is expected to continue on the IPBES, with a final decision on its establishment and institutional arrangements expected at the Governing Council Session in 2011.

Clearly there is support for the principle of strong science helping to inform decision-makers. The concern we have is that just as conventions are beginning to cluster, there is a potential risk of fragmentation of the scientific aspect. With this in mind, some governments have suggested that perhaps there needs to be an overarching Intergovernmental Panel on Sustainable Development as an umbrella body under which other intergovernmental panels would be institutionalized. Ideally, the placing of this panel would be within a strengthened UNEP.

UN system-wide coherence

At the moment, two main bodies undertake interagency work on biodiversity, namely: the Joint Liaison Group of the Biodiversity Conventions and the UN Environment Management Group (UNEMG), which is coordinated by UNEP.

The Joint Liaison Group was established in 2002, to bring together the seven biodiversity related conventions: CBD; Convention on Migratory Species (CMS); Convention on International Trade in Endangered Species (CITES); International Treaty on Plant Genetic Resources for Food and Agriculture (IT PGFR); Ramsar Convention on Wetlands; and World Heritage Convention (WHC). The group, which meets regularly, is mandated to explore opportunities for synergistic activities, increased coordination and to exchange information. In this regard its mandate has been curtailed to mere coordination issues, and participation is limited to treaty bodies and therefore does not include the core UN system organizations working on biodiversity. However, the UN system organizations are active participants in the Heads of Agency Task Force on the 2010 Biodiversity Target.

The Chief Executive Board (CEB) for Coordination brings together on a regular basis the executive heads of the organizations of the UN system, under the chairmanship of the Secretary-General of the UN. Following the 2002 WSSD, the board established three interagency processes – UN Oceans, UN Water and UN Energy – as mechanisms to coordinate within the UN system on these key areas. UN Water for example is an interagency that includes 26 members from the UN system as well as external partners representing civil society and various organizations. While a welcomed effort by the UN system, developing countries were displeased that these bodies were set up without input and direction from member states. Moreover, it is a likely reason for the UN system's climate change response following a more programmatic and less institutional process.

Diplomacy aside, further consideration needs to be given to establishing UN Biodiversity to bring together the interagency work of the UN system and outside partners, particularly since the current interagency work is placed too far away from the UN Center in New York. UN Biodiversity would replace the Issue Management Group under the UN Environmental Management Group, elevating interagency status under the CEB, with direct access to the UN Development Group and thus the UN's operational structures at national and regional levels, including through the Country Cooperation Frameworks and UN Development Assistance Frameworks.

Global biodiversity indicators

Chapter 40 of Agenda 21 called upon countries, international governments and non-governmental organizations (NGOs) to develop indicators of sustainable development that could provide a solid basis for decision-making at all levels. Furthermore, Agenda 21 called for the harmonization of efforts to develop such indicators. However, indicators for sustainable development have been a difficult topic to address at the intergovernmental level. In many places the data systems that collect information have at times been incompatible, or in the Least Developed Countries (LDCs), non-existent. There has been much work in this area since the Rio Summit but often it fails to form a coherent framework because of a difference in viewpoints relating to whether there needs to be a common set of indicators or a basket of indicators that governments can pick and choose from. We believe that the basket approach is the best way forward and may in time deliver a core set of indicators. Current work linking indicators and policy decisions should not be halted because of disagreements like this. The UN CSD failed in 1998 because of such disagreements. Any discussion of policy should be supported by a basket of indicators. The CBD COP in 2004 identified a suite of 17 headline indicators from the seven focal areas for assessing progress towards, and communicating, the 2010 target at a global level. This should continue and be developed further (CBD, 2010).

Clustering and joint work programmes

In JPOI (Chapter 3, paragraph 44 (c)), the international community decided that in order to meet the 2010 target 'effective synergies between the Convention and other MEAs, among other things, through the development of joint plans and programmes, with due regard to their respective mandates, regarding common responsibilities and concerns' would need to be encouraged. In this regard, the CBD COP have taken a number of measures in this direction, particularly in relation to climate change, but on most occasions making progress and getting support from other Multilateral Environmental Agreement COPs has proven particularly difficult and often frustrating.

In 2010, the idea of clustering conventions covering similar areas took a major step forward with the convening of a simultaneous, extraordinary meeting of the COP to the Basel, Rotterdam and Stockholm conventions in Bali, Indonesia. It was hoped that this historic convening of the three independent treaty conferences simultaneously would mark a step forward for the stalled IEG discussions on the coherence and clustering of Multilateral Environmental Agreements at the global level. In preparation for the 14th Session of the CDB Subsidiary Body for Scientific and Technological Advice (SBSTA), the Secretariat prepared a background paper to stimulate Party thinking on a more coordinated approach to the relationship between biodiversity, climate change and desertification. The Secretariat Paper outlined four programmatic areas of cooperation, namely: integrated and coordinated national planning linking biodiversity, climate change and land degradation; addressing the common drivers of biodiversity loss, climate change and land degradation and desertification; understanding, monitoring, assessing and reporting on the interlinkages between biodiversity, climate change, land degradation and desertification and sustainable development; and promoting a favourable enabling environment. What is clearly emerging as the major thematic area for joint programmatic work, coordination and decision-making is the issue of Reducing Emissions from Deforestation and Forest Degradation (REDD) in developing countries. All three conventions have worked in this area and a common meeting of the three conventions to discuss and agree a common programme of work would go a long way to address fragmentation.

Similarly, the Biodiversity Liaison Group[1] meets regularly to explore opportunities for synergistic activities and increased coordination, and to exchange information. This is at the inter-agency level but the next move would be to host a common meeting of the five Biodiversity Conventions.

Life in Balance

The challenges are great to secure and control the reduction of the loss of biodiversity and put in place the robust intergovernmental systems that are needed to address biodiversity and ecosystem insecurity. The longer we leave this the more likely it is we will see major ecosystem failures. Can we continue to be irresponsible when we know the potential for disaster? Especially for a disaster that will affect the poorest first. This is not acceptable.

Note

1 The Biodiversity Liaison Group consists of the following: CBD, CMS, CITES, IT PGFR, Ramsar Convention on Wetlands and WHC.

References

Alcamo, J. (2010) 'UN-backed study reveals rapid biodiversity loss despite pledge to curb the decline', UN News Centre, www.un.org/apps/news/story.asp?NewsID=34557&Cr=biodiversity&Cr1=

CBD (Convention on Biological Diversity) (2010) 'Faced with biodiversity crisis, a new vision is urgently required', press release, 25 May 2010

Dodds, F., Howell, M., Onestini, M. and Strauss, M. (2007) 'Negotiating and implementing Multilateral Environmental Agreements, United Nations Environment Management Group', www.unemg.org

GBO3 (Global Biodiversity Outlook 3) (2010) 'United Nations Environment Programme', press release, www.cbd.int/doc/press/2010/pr-2010-05-10-gbo3-en.pdf

INTGLIM (International NGO Task Group on Legal and Institutional Matters) (1997) *Renewing the Spirit of Rio: The CSD, Agenda 21, and Earth Summit +5*, International NGO Task Group on Legal and Institutional Matters, http://habitat.igc.org/csd-97/pt5-10.htm

Najam, A., Papa, M. and Taiyab, N. (2006) *Global Environment Governance: A Reform Agenda*, International Institute for Sustainable Development, www.iisd.org/pdf/2006/geg.pdf

Rockström, J., Steffen, W., Noone, K., Persson, Å., Chapin, F. S., Lambin, E. F., Lenton, T. M., Scheffer, M., Folke, C., Schellnhuber, H. J., Nykvist, B., de Wit, C. A., Hughes, T., van der Leeuw, S., Rodhe, H., Sörlin, S., Snyder, P. K., Costanza, R., Svedin, U., Falkenmark, M., Karlberg, L., Corell, R. W., Fabry, V. J., Hansen, J., Walker, B., Liverman, D., Richardson, K., Crutzen, P. and Foley, J. A. (2009) 'A safe operating space for humanity', *Nature*, vol 461, no 7263, pp472–475

UN (United Nations) (2002) 'Johannesburg Plan of Implementation of the World Summit on Sustainable Development', United Nations

UNEP (United Nations Environment Programme) (2007) 'Negotiating and implementing MEAs, UNEP', United Nations Environment Programme, Nairobi

17

The Economics of Ecosystems and Biodiversity (TEEB) and Its Importance for Governing Biodiversity

Pavan Sukhdev and Christoph Schröter-Schlaack[1]

Why Is Biodiversity Neglected in Decision-making?

Ecosystems, biodiversity and natural resources underpin economies, societies and individual well-being. The values of its myriad benefits are, however, often overlooked or poorly understood. They are rarely taken fully into account through economic signals in markets, or in day-to-day decisions by business and citizens, nor indeed reflected adequately in the accounts of society.

Table 17.1 *Definitions of biodiversity strata*

Biodiversity Strata	Quality dimensions	Quantity dimensions	Ecosystem Services (exemplarily)
Ecosystem	Variety	Extent	• Recreation • Water regulation • Carbon storage
Species	Diversity	Abundance	• Food, fibre, fuelwood • Design inspiration • Pollination
Genes	Variability	Population	• Medicine discovery • Disease regulation • Adaptive capacity

In recent decades, nearly all countries have adopted targets and rules to conserve species and habitats and to protect the environment against pollution and other damaging activities. Policies and measures that have positively affected biodiversity and ecosystem services have taken a wide variety of forms, from regulatory instruments – such as the protection of critically important habitats – to reforming environmentally harmful subsidies to incentive-based mechanisms such as payments for ecosystem services or measures based on the 'polluter pays' and 'full cost recovery' principles.

Despite this progress, the scale of the global biodiversity crisis shows that current policies are simply not enough to tackle the problem efficiently. Some of the reasons are only too familiar to policy-makers, such as lack of financial resources, lack of capacity, information and/or expertise, overlapping mandates and weak enforcement. But there are also more fundamental economic obstacles in this policy field that we need to understand to make meaningful progress.

A root cause of the systematic neglect of ecosystems and biodiversity is their economic invisibility, due to their nature as a public good – and often a global public good. Biodiversity benefits take many forms and are widespread, which makes it difficult to 'capture' value and ensure that beneficiaries pay for them. For example, a forest provides local benefits to local people (timber, food and other products); the forest ecosystem mediates water flows and provides regional climate stability; and forests are also globally important because they sustain biodiversity and act as long-term carbon sinks. So far, existing markets and market prices only capture some ecosystem services (for example, ecotourism, water supply). More commonly, individuals and businesses can use what biodiversity provides without having to pay for it, and those providing the service often do not get due recompense. Although the costs of conservation and restoration are paid immediately, often at local level, many benefits occur in the future. For example, creating a protected area to save an endangered species can cause short-term losses to user groups, which may lead us to give little or no weight to the possible long-term benefits (such as discovery of medicinal cures from such species).

What is more, uncertainty regarding potential future benefits is matched by ignorance about the risks of inaction. We know too little about why each species is important, for example, what its role in the food web is and what could happen if it is lost. At a larger scale, we do not know the 'tipping points' of different ecosystems and, for example, their role in the hydrological cycle. Uncertainties lead policy-makers to hesitate: spending money on policies with measurable and market-priced returns and potential for taxable revenues seems preferable to spending on policies with less visible outcomes or non-market-priced welfare benefits. And finally, the deterioration of ecosystem services and biodiversity often occurs gradually. Marginal impacts of individual and local actions can add up to severe damage at the global scale. For example, small-scale assessments of individual development projects (such as forest clearance for agriculture or housing) can indicate a positive cost–benefit ratio but cumulative impacts in terms of deforestation and habitat fragmentation can be far higher.

Decisions regarding management of biodiversity involve trade-offs: if we want to keep ecosystem services, we often have to give something up in return. Currently, where trade-offs have to be made between biodiversity conservation and other policy areas (such as agriculture, industry, transport, energy), the lack of compelling economic arguments means that decisions very often go against biodiversity conservation.

Economic Information to Improve Public Policies and Private Decisions Affecting Biodiversity

Though governments and public authorities are responsible for setting policy, a whole series of other groups (industry and business, consumers, landowners, non-governmental organizations (NGOs), lobbyists, indigenous people, and so on) also make decisions that affect the natural environment. The challenge is to identify all relevant actors, mobilize 'leaders' and ensure that they have the necessary information and encouragement to make the difference.

There is a compelling rationale for governments to lead efforts to safeguard ecosystem services and biodiversity. Public environmental policy needs to be based on moral values (concern for human well-being), intrinsic values (not letting species go extinct) and good stewardship, while taking economic considerations into account. Natural capital – the economic reflection of ecosystems and biodiversity and their welfare benefit flows – forms a significant part of national public wealth. Public policy must therefore recognize and seek to optimize its returns to society, and new policy responses should be shaped and guided to reduce ongoing losses of natural capital and to invest in the healthy functioning of ecosystems to deliver benefits into the future.

Private actors (businesses and consumers) have a growing role to play in choices that affect our natural capital. However, a strong policy framework is needed to ensure that decisions are efficient – in other words, ensuring that society gets the most from its scarce biodiversity resources – and equitable, such as ensuring benefits of biodiversity are distributed fairly. Appropriate regulation provides the context in which private markets for ecosystem services can evolve as well as mechanisms to monitor their effectiveness.

Focusing on the services provided by biodiversity and ecosystems is critical to overcome their traditional neglect. The Millennium Ecosystem Assessment (MEA, 2005) paved the way for indicators to show the status of ecosystem services. The transition from acknowledging services to valuing them may seem a small step but it is a huge step towards raising awareness. In doing so, one can demonstrate that biodiversity and ecosystem services have value, not only in the narrow sense of goods and services in the marketplace but also – and more importantly – because they are essential for our survival and well-being. This is the case even if markets do not exist or if these values are not expressed in monetary terms: values can also be based on qualitative assessments. Indeed, values can be recognized by societies and communities and acted upon, and yet may not be economically demonstrated, let alone captured. What we actually measure in monetized form is very often only a partial reflection or a share of the total value of ecosystem services and biodiversity (TEEB, 2008, p33).

Using economic values in the choice and design of policy instruments can help overcome the systematic bias in decision-making by demonstrating the existence of economic values beyond those that are market-priced and evidenced (for example, beyond manufactured capital to natural capital, beyond present to future benefits/costs and different resource types). Furthermore, even if biodiversity benefits are multifaceted and diffuse, economic valuation helps to subsume or aggregate benefits within certain broader values (such as for forests) and this in turn may help to create new mechanisms and markets where none previously existed. The recently created markets for greenhouse gas (GHG) emissions are powerful examples from climate policy of what can be achieved where market-based approaches are developed for environmental goods within a strong policy framework. Finally, economic values help to make future benefits visible, rather than simply relying on today's costs (for example, by identifying option values of plants from tropical forests relevant for pharmaceutical products, or the potential of tourism).

Successful biodiversity policies are often restricted to a small number of countries, because they are unknown or poorly understood beyond these countries. Economics can highlight that there are policies that already work well, deliver more benefits than costs and are effective and efficient. The Reducing Emissions from Deforestation and Forest Degradation (REDD) scheme and its variants such as REDD+[2] have already stimulated broader interest in payment for ecosystem services (PES). Several countries and organizations have collated case studies on REDD design and implementation that can be useful for other countries and applications (Parker et al, 2009). Other examples of approaches that could be used more widely for biodiversity objectives include instruments based on the polluter-pays principle, such as taxes, fees or permit trading and green public procurement.

Moreover, economic analysis can help existing instruments work better. Using assessment tools to measure and compare the efficiency and cost-effectiveness of existing policies can ensure that instruments reach their full potential. Assessment provides ongoing opportunities to review and improve policy design, adjust targets and thresholds and make the positive effects of protection visible (such as for protected areas). The process increases transparency and can contribute to acceptance of biodiversity policies by stakeholders. Economic assessment can make explicit the damage caused by harmful subsidies, for example, subsidies for housing that encourage land conversion and urban sprawl in natural areas and fisheries or agricultural subsidies that are harmful to biodiversity and ecosystems (UNEP, 2008; OECD, 2009).

Economic information allows policy-makers to simultaneously address poverty issues and social goals if the distribution of costs and benefits to different groups in society is included in the analysis. Such analysis can highlight the importance biodiversity and ecosystem services have for poorer segments of the population in many countries. When designed accordingly, biodiversity policies can contribute to alleviating poverty (Tallis et al, 2008).

Guiding Principles for Governing Biodiversity

Measuring what we manage: Information tools for decision-makers

Unlike physical and financial capital, natural capital has no dedicated systems of measurement, monitoring and reporting. This is astonishing given its importance for jobs and mainstream economic sectors, as well as its contribution to future economic development. For instance, we have only scratched the surface of what natural processes and genetic resources have to offer. As part of good governance, decision-making that affects people and the use of public funds needs to be objective, balanced and transparent. Access to the right information at the right time is fundamental to coherent policy trade-offs. Better understanding and quantitative measurement of biodiversity and ecosystem values to support integrated policy assessments are a core part of the long-term solution.

The first key need is to improve and systematically use science-based indicators to measure impacts and progress and alert us to possible 'tipping points' (sudden ecosystem collapse). Specific ecosystem service indicators are needed alongside existing biodiversity tools. Another key need is to extend national income accounts and other accounting systems to take the value of nature into account and monitor how natural assets depreciate or grow in value with appropriate investments. New approaches to macro-economic measurement must cover the value of ecosystem services, especially to those who depend on them most, such as poor rural and forest-dwelling communities.

Addressing the right actors and balancing diverse interests

Biodiversity is the ultimate cross-cutting issue and several policy fields have significant implications for biodiversity (transportation, trade, land use, regional planning, and so on). Such policies can have negative impacts on biodiversity or be designed to promote positive synergies. Even within single sectors, there is a broad range of different stakeholders and interests. Production patterns can vary from environmentally sensitive to high impact. Within agriculture, for example, eco-farming is associated with sustainable land-use practices and mitigation of soil depletion or erosion, whereas industrialized farming involves monocultures and intensive use of fertilizers and pesticides.

Additional challenges arise where policy-making involves several governmental levels, such as global negotiation rounds or supranational organizations, national policy-makers, regional administration or local interest groups. Many international agreements and mechanisms are in place to streamline cooperation across boundaries. To improve water resource management, for example, more than 80 special commissions with three or more neighbours have been established in 62 international river basins (Dombrowsky, 2008).

However, policy-makers can build on the high number of treaties that target the protection of species, habitats, genetic diversity or biodiversity as a whole, such as the Convention on Biological Diversity (CBD), the Convention on Wetlands or the World Heritage Convention (WHC), to name but a few. In parallel, however, adopting the ecosystem services approach may necessitate amendments to international conventions and standards in other policy sectors. For example, current World Trade Organization (WTO) rules prohibit the introduction of certain environmental standards (one example is in respect to timber) as they would violate free trade principles.

Mechanisms to ensure policy coordination and coherence between different sectors and levels of government are therefore essential, both within and between countries. Spatial planning is an important part of this equation. A large amount of environmental decision-making takes place close to the ground (for example, permitting, inspection, planning decisions and enforcement), which means that local administrations and actors need to be aware, involved and adequately resourced.

Paying attention to the cultural and institutional context

A country's cultural context (religious norms or morality, level of civil society engagement) and institutional context (laws, regulations, traditions) can provide useful entry points for biodiversity conservation. Policy options may be easier to implement and enforce when they fit easily into existing regulations and do not need substantial legislative changes or reallocation of decision-making power. Establishing a protected area or restricting use of a certain resource can be easier if backed by religious norms. Market-based tools to manage ecosystem services may be more easily accepted in countries that use markets for pollution control or nature protection than in regions relying on traditional regulatory norms or structures.

As in any policy area, new instruments and measures to protect biodiversity can face difficulties not only when being negotiated but also in day-to-day implementation and enforcement. Good design, good communication and goodwill are particularly important to boost compliance with environmental policy instruments that need backing from affected stakeholders to be fully effective. For example, payment schemes to reward biodiversity-friendly agricultural practices will only work well if people fully understand the scheme and do not face other obstacles.

'Windows of opportunity' can open in response to increased awareness of environmental problems; increasing public concern over the ozone led to the Montreal Protocol and the same concern over climate change produced the REDD mechanism, which has great potential for broader application to biodiversity-related issues. Current crises (such as food prices, oil prices, credit) could provide new opportunities to phase out expensive subsidies harmful to biodiversity, for example, in agriculture or fisheries. Policy windows can also result from reaction to catastrophes, for example, oil spills or natural disasters.

One country's move is another country's (window of) opportunity to follow. Political 'champions' who propel a new problem up the policy agenda and offer innovative solutions (for example, PES in Costa Rica, REDD in Guyana) can catalyse progress at a regional or global level. Sharing information about success stories (as TEEB is doing, for example) is a practical way to learn from experience elsewhere and develop solutions appropriate to national needs and priorities.

Taking property rights, fairness and equity into account

New strategies and tools for protecting biodiversity and sustaining ecosystem services often involve changes in rights to manage, access or use resources ('property rights'). The distributional implications of policy change, particularly for vulnerable groups and indigenous people, require up-front identification and consultation throughout the policy development process.

At least three arguments support consideration of property rights and distributional impacts as an integral part of policy development. Firstly, equity considerations – such as fairness in addressing changes of rights between individuals, groups, communities and even generations – are an important policy goal in most countries. Secondly, taking distributional issues into account makes it much more feasible to achieve other goals when addressing biodiversity loss, particularly related to poverty alleviation and the Millennium Development Goals (MDGs) (UNDP, 2010). Lastly, there are almost always winners and losers from policy change and, in most cases, loser groups will oppose the policy measures. If distributional aspects are considered when designing policies, the chances of successful implementation can be improved.

What complicates matters for the policy-maker is that different rights are often held by different people or groups in society. A forest might be owned by the state but local people might have a right to use some of its products. Rights for water coming from this area might be held by third parties and international companies might hold concessions for deforestation. This legal and historic complexity needs to be considered when adjusting or introducing policies for ecosystem services and biodiversity. Distributional issues specifically arise where benefits of ecosystem conservation go beyond the local level. For example, restricting land use upstream is often necessary to maintain freshwater provision at adequate levels and quality downstream. Where distributional impacts are perceived as unfair, compensation may be necessary to ensure full implementation of selected policies.

Decision-making today also affects tomorrow's societies: the species we commit to extinction are clearly not available to future generations. If ecosystems can no longer provide important regulating services, future generations will have to provide for them in a different manner. This has serious ethical implications, further compounded by the use of high discount rates. For example, a social discount rate of around 4 per cent (commonplace in the literature) implies a trade-off in which our grandchildren (50 years from now) have a right to only one seventh of what we use today (TEEB, 2008, p30).

Responding to the Value of Nature

Faced with the growing threat from climate change, governments have started focusing on the need to move towards a low-carbon economy, an economy that minimizes GHG emissions. There is a need and an opportunity to take this concept a step further towards a truly resource efficient economy. An economy that sends out signals that reflect the many values of nature, from the provision of food, raw materials and access to clean water, all the way up to recreation, inspiration and a sense of cultural and spiritual identity; an economy that makes the best use of its biodiversity, ecosystems and resources without compromising their sustainability; an economy supported by societies that value their natural capital.

It is hard to think of any other asset class where we would tolerate its loss without asking ourselves what we risk losing and why. The more that we ask these questions, the more uncomfortable we become with the current situation where nature is being lost at an alarming rate. We realize that we often fail to ask the big questions about what ecosystem services and biodiversity provide and their value or worth to different groups of people, including the poorest, across the globe and over time.

Building momentum for the transition to a resource efficient economy – in the broadest sense – calls for international cooperation, partnerships and communication. Every country is different and will need to tailor its responses to the national context. However, all may stand to gain – countries, businesses, and people on the ground – by sharing ideas, experience and capacity. Policy champions can lead this process and use windows of opportunity to forge a new consensus to protect biodiversity and ecosystems and their flows of services. We hope that the TEEB report suite contributes to this new momentum.

The goal of *The Economics of Ecosystems and Biodiversity: Ecological and Economic Foundations* (Kumar, 2010) is to provide the conceptual foundation to link economics and ecology and to posit a paradigm of the relationship between biodiversity and ecosystem services. This aspect of the study tackles the challenges of valuing ecosystem services, as well as issues related to economic discounting. It aims to quantify the costs of inaction and examine the macro-economic dimension of ecosystem services loss. This information will focus on improving our understanding of the economic costs of biodiversity loss and ecosystem degradation. The process is bringing scientists and economists together to provide the analysis and tools required in order to be able to create a robust methodological framework enabling the decision-makers at different levels to do economic analysis of ecosystem services and biodiversity.

The Economics of Ecosystems and Biodiversity in National and International Policy (ten Brink, 2011) is a contribution to the call by an increasing number of policy-makers for ways to approach this multifaceted challenge. It shows that the accumulated policy experience is plentiful and provides a broad range of solutions. At present these are mainly carried out in isolation, creating pockets but also important starting points. The creativity and vision of international and national policy-makers is now in demand to design coherent policy frameworks that systematically respond to the value of nature. These can open up new opportunities to address poverty, development and growth. At the same time, the act of making values visible through well-designed policies will empower consumers, business, communities and citizens to make much more informed choices and thus to contribute to this transition in their daily decisions.

The Economics of Ecosystems and Biodiversity in Local and Regional Policy (Wittner and Gundimedia, 2011) complements, translates and adapts the findings of the TEEB Foundations and the messages of *The Economics of Ecosystems and Biodiversity in National and International Policy* to the local/regional policy levels. This report will present tools and applications in various fields such as spatial planning, urban management, natural resource management and protected areas. To this end, it will draw on poignant case studies that illustrate the policy uptake and the actual impact of different tools.

The Economics of Ecosystems and Biodiversity in Business and Enterprise (Bishop, 2011) acknowledges the huge role business and enterprise have to play in how we manage, safeguard and invest in our natural capital. Aimed squarely at this sector, the report will provide practical guidance on the issues and the opportunities created by the inclusion in mainstream business practices of ecosystem- and biodiversity-related considerations. This report is for a wide array of enterprises, including those with direct impacts on ecosystems and biodiversity such as mining, oil and gas and infrastructure; for those businesses that depend on healthy ecosystems and biodiversity for production, such as agriculture and fisheries; for industry sectors that finance and undergird economic activity and growth, such as banks and asset managers, as well as insurance and business services; and for businesses that are selling ecosystem services or biodiversity-related products such as ecotourism, ecoagriculture and biocarbon.

Making a truly resource efficient economy a reality will require tremendous effort and international cooperation, but the existing evidence shows that it will undoubtedly be worthwhile. The future is in our hands and we have the potential to make the outlook much more positive. Acknowledging and quantifying the value of nature's flows means that decisions can be made now that will reap sustained environmental, social and economic benefits far into the future, supporting future generations as well as our own.

Notes

1 As this chapter builds on *The Economics of Ecosystems and Biodiversity in National Policy* (ten Brink, 2011), the authors are grateful to its editor, Patrick ten Brink, and the many people involved in his working team.

2 REDD+ means reducing emissions from deforestation and forest degradation in developing countries, and the role of conservation, sustainable management of forests and enhancement of forest carbon stocks in developing countries according to the Bali Action Plan.

References

Bishop, J. (ed) (2011) *The Economics of Ecosystems and Biodiversity in Business and Enterprise*, Earthscan, London (forthcoming)

Dombrowsky, I. (2008) 'Integration in the management of international waters: Economic perspectives on a global policy discourse', *Global Governance*, vol 14, pp455–477

Kumar, P. (ed) (2010) *The Economics of Ecosystems and Biodiversity: Ecological and Economic Foundations*, Earthscan, London

MEA (Millennium Ecosystem Assessment) (2005) *Ecosystems and Human Well-being: Biodiversity Synthesis*, Island Press, Washington, DC

OECD (Organisation for Economic Co-operation and Development) (2009) *Agricultural Policies in OECD Countries: Monitoring and Evaluation*, Organisation for Economic Co-operation and Development, Paris

Parker, C., Mitchell, A., Trivedi, M. and Mardas, N. (2009) *The Little REDD+ Book*, Global Canopy Programme, Oxford

Tallis, H., Kareiva, P., Marvier, M. and Chang, A. (2008) 'An ecosystem services framework to support both practical conservation and economic development', *PNAS*, vol 105, no 28, pp9457–9464

TEEB (The Economics of Ecosystems and Biodiversity) (2008) *TEEB An Interim Report*, www.teebweb.org

ten Brink, P. (ed) (2011) *The Economics of Ecosystems and Biodiversity in National and International Policy*, Earthscan, London (forthcoming)

Wittmer, H. and Gundimedia, H. (2011) *The Economics of Ecosystems and Biodiversity in Local and Regional Policy*, Earthscan, London (forthcoming)

UNDP (United Nations Development Programme) (2010) 'Millennium Development Goals', www.undp.org/mdg/

UNEP (United Nations Environmental Programme) (2008) *Fisheries Subsidies: A Critical Issue for Trade and Sustainable Development at the WTO: An Introductory Guide*, United Nations Environmental Programme, Geneva

18

Traditional Knowledge in Global Policy-making: Conservation and Sustainable Use of Biological Diversity

John Scott

Introduction

Traditional knowledge is a tool, proven by the test of time that has allowed many ecosystem-based people to cope with change and prosper over millennia. It has contributed directly to their resilience and the flourishing of both cultural and biological diversity. The extent to which the mainstream can access and learn from such knowledge, innovations and practices, will directly contribute to humanity's ability to cope with such pressing global issues as climate change, biological diversity loss and the unsustainable use of our limited resources.

Most indigenous and local communities are situated in areas where the vast majority of the world's biological diversity, including genetic diversity, is found. Many of them have cultivated and used biological diversity in a sustainable way for thousands of years. Some of their practices have been proven to enhance and promote biodiversity at the local level and aid in maintaining healthy ecosystems. However, the contribution of indigenous and local communities to the conservation and sustainable use of biological diversity goes far beyond their role as natural resource managers. Their skills and techniques provide valuable information to the global community and a useful model for biodiversity policies. Furthermore, as on-site communities with extensive knowledge of local environments, indigenous and local communities are most directly involved with *in situ* conservation and sustainable use. Since indigenous peoples are an integral part of the ecosystem they manage, the best guarantee for the survival of nature (environmental sustainability) is the survival of this knowledge and of the holders of this knowledge.

This chapter explores the role of traditional knowledge in conservation and sustainable use within the context of global policy-making and the Convention on Biological Diversity (CBD).

What Is Traditional Knowledge?

Traditional knowledge refers to the knowledge, innovations and practices of indigenous and local communities around the world. Developed from experience gained over the centuries and adapted to the local culture and environment, traditional knowledge is transmitted orally from generation to generation. It tends to be collectively owned and takes the form of stories, songs, folklore, proverbs, cultural values, beliefs, rituals, community laws, local language and agricultural practices, including the development of plant species and animal breeds. Sometimes it is referred to as an oral tradition for it is practised, sung, danced, painted, carved, chanted and performed down through millennia. Traditional knowledge is mainly of a practical nature, particularly in such fields as agriculture, animal husbandry, fisheries, health, horticulture, forestry and environmental management in general.

There is today a growing appreciation of the value of traditional knowledge. It is valuable not only to those who depend on it in their daily lives, but also to modern industry and agriculture. Many widely used products, such as plant-based medicines, health products and cosmetics, are derived from traditional knowledge. Other valuable products based on traditional knowledge include agricultural and non-wood forest products as well as handicraft.

There are many worthy definitions of traditional knowledge used throughout academia. For our purposes, the meeting held under the CBD of technical and legal experts on traditional knowledge (associated with genetic resources in the context of the international regime on access and benefit sharing), proposed the following agreed characteristics that may be useful to consider in this discussion:

- a link to a particular culture or people and/or place knowledge is created in a cultural context;
- a long period of development, often through an oral tradition, by unspecified creators;
- a dynamic and evolving nature;
- existence in codified or non-codified (oral) forms;
- passed on from generation to generation – intergenerational in nature;
- local in nature and often imbedded in local languages;
- unique manner of creation – (innovations and practices); and
- it may be difficult to identify original creators (hence mainly collective ownership).

Value of Traditional Knowledge to the Holders of the Knowledge

The value of traditional knowledge to the holders and originators of the knowledge is apparent, given that it is the embodiment of a way of life, including culturally appropriate decision-making processes. Traditional knowledge is a holistic knowledge system that provides a framework through which knowledge is created, tested and passed on, and is the means through which the peoples and communities who hold the bodies of knowledge make decisions on day-to-day matters and their immediate and long-term futures. Many indigenous peoples have traditions that, in fact, require long-term planning. For instance, indigenous peoples in North America talk in terms of 'the seventh generation'. In the occidental world, where long-term planning rarely extends beyond political cycles of three to five years, there are important lessons to learn from taking into account the seventh generation.

The broad nature of traditional knowledge also provides practical illustration of the importance of traditional knowledge to those who hold it. Traditional knowledge provides extremely important information in regards to hydrology, medicines, food production and security, agricultural systems, animal husbandry and many other crucial matters, and as such it is integrally tied to the viability and resilience of a people or community.

In regard to health, for example, as many as one third of the world's population do not have access to drugs developed through Western scientific methods, which places a huge reliance on traditional medicines that are local and easily accessible (Zhang, 2004). Furthermore, as much as 80 per cent of the world's population continues to rely on traditional medicines and health practitioners for their primary medical needs (RAFI, 1995), which in turn heavily relies on an intimate knowledge of the environment and ecologies within which a people or communities live.

Similar examples are available for food production and security. Between half and two thirds of the world's population – particularly indigenous peoples and local communities – are almost wholly dependent on their own food production systems (RAFI, 1995), which incorporates traditional knowledge on such diverse subject matter as cropping methods, seed production, selection and storage, cultivar development through experimentation, knowledge of animal behaviour and animal husbandry as well as plant diversity, among many other factors.

While it is useful to broadly recognize the value of traditional knowledge, to speak of the 'intrinsic value' of traditional knowledge to indigenous peoples and local communities does a disservice to the real and daily importance of traditional knowledge to its holders. Bodies of traditional knowledge are not merely 'things' or 'objects' or 'property', and should therefore not be objectified. Rather, bodies of traditional knowledge are systems of knowing and ways of life that peoples and communities are situated within and, as such, are beyond value for those who live within them. To speak of 'value' can lead to reductionism and greatly underemphasizes the true importance of traditional knowledge to its holders and originators. With this in mind, it is possible to gain a truer understanding of the grave consequences to societies and cultures that the loss of traditional knowledge can have.

The Value of Traditional Knowledge for Conservation and Sustainable Use

Indigenous peoples and local communities have an encyclopaedic knowledge of their local environment that has been handed down orally from generation to generation, sometimes spanning millennia. Many of their practices, not only preserved, but increased biodiversity on their traditional lands and waters.

Such cultures have developed ways to maintain, encourage and even increase biological diversity on their traditional territories through such diverse practices as fire management and spot farming in the Wet Tropics of North Eastern Australia to traditional water management practices in the Middle East. Traditional fire management practices in the Wet Tropics of Far Northern Queensland, Australia, have increased the biological diversity in jungle areas, by encouraging grazing animals, such as kangaroos and wallabies into the open grassy clearings in the rainforest, created by burnings. Also, small marsupials such as the northern bettong are dependent on fungi and mushrooms that grow on the edges of clearings of this type and without these practices, they soon disappear. Such practices are now being considered and implemented on a broad scale, often in partnership with indigenous peoples, by protected area authorities. Fire management is such a long established practice in the traditional Australian landscape that many plant seeds, such as wattle bush, have evolved in such a way that they will not germinate until they are exposed to fire. However, hot wildfires will kill them. Their germination requires what is referred to as a 'cool fire', which is the result of regular and controlled burning of the bush without excessive build up of underbrush.

The CBD very much equates sustainable use with conservation. Article 10(c) of the CBD calls on parties to protect and encourage customary use of biological resources in accordance with traditional cultural practices that are compatible with conservation and sustainable use.

However, advancing and implementing Article 10(c) requires consideration of the common characteristics of 'customary use of biological resources in accordance with traditional cultural practices'. The customary use of biological resources is but one aspect of the knowledge-practice-belief complex that makes up the traditional ecological knowledge of indigenous and local communities. An analysis of traditional knowledge systems shows that they tend to include three major components: observation of ecological conditions, cultural practices governing resources activity, and belief about how people fit into the ecosystem (Berkes, 2000). The CBD acknowledges the interconnectedness of traditional knowledge and customary practice in the preamble and in Article 8(j) by referring to 'traditional knowledge, innovations and practices' as related cultural realities that are equally deserving of recognition and protection.

Understanding customary use as the application of traditional knowledge

Given the mutually dependent relationship between traditional knowledge and customary practice, no meaningful discussion of how to advance customary sustainable use initiatives can be undertaken without an awareness of the greater discussion surrounding the need to protect traditional knowledge. For the purposes of this chapter, 'customary use of biological resources in accordance with traditional cultural practices' will be taken to broadly refer to uses of biological resources pursuant to the customary application of the traditional ecological knowledge of indigenous and local communities.

The importance of fundamental human rights to customary sustainable use

Because customary practices are intimately tied to the culture, traditions and beliefs of indigenous and local communities, encouraging customary sustainable practices will necessarily require recognizing and encouraging the cultural uniqueness and diversity of indigenous peoples and local communities. Recognition of the fundamental human rights of indigenous peoples is therefore crucial to the viability and vitality of customary sustainable use practices. Indigenous and local communities will only be able to contribute to conservation initiatives if their fundamental rights are secure and they are able to engage in their customary practices freely and openly.

The United Nations Declaration on the Rights of Indigenous Peoples (UNDRIP) provides a framework for respecting the fundamental human rights of indigenous peoples and provides a legal basis for protecting and encouraging customary sustainable use. The preamble of the declaration reaffirms that, while indigenous peoples are entitled to all the human rights recognized in international law, they also possess collective rights, which are indispensable for their existence and well-being and integral to their development as peoples. These collective rights include the right to practice and revitalize their cultural traditions and customs, the right to be secure in the enjoyment of their own means of subsistence and development, and the right to engage freely in all their traditional and other economic activities. The preamble of the declaration recognizes that respect for traditional practices, cultures and indigenous knowledge contributes to sustainable and equitable development, and proper management of the environment.

Customary sustainable use practices as a characteristic of indigenous and local communities

Despite wide acknowledgement of the important role to be played by customary practices in the preservation of biological diversity, care must be taken not to idealize and distort the realities of indigenous traditional knowledge and customary use of resources. For one, not all traditional societies have always lived in harmony with nature (Berkes, 1999) and some of the practices of indigenous and local communities still living traditional lifestyles may be ecologically unsound (Gadgill, 1993) However, relatively sedentary societies relying on fishing, horticulture, subsistence agriculture, and/or hunting and gathering are very likely to have accumulated a substantial amount of knowledge relevant to sustainable resource use and management, and the conservation of biodiversity in their traditional territory. Societies faced with limited resources and limited opportunities to relocate, tend to develop self-regulatory mechanisms to avoid overexploitation, including the accumulation of knowledge about the ecological services and natural resources provided by species and landscapes. Thus, traditional knowledge and practices relevant to conservation are often characteristic features of societies with a long history of continuous use of a given resource base.

A long history of interaction between a community and an ecosystem does not only have socio-cultural consequences, it also has ecological consequences. Over time, as human communities interact closely with their local environments and adapt to fill ecological niches, local environments are modified and local ecological processes adjust to human influences. Indeed, a strong correlation has been found between areas of high cultural diversity, which tend to be characterized by relatively large numbers of indigenous communities and areas of high biological diversity (Oviedo, 2000). Indigenous and local communities can be understood as 'ecosystem peoples' who have co-evolved with their environment, and indigenous cultures can be understood as 'cultures of habitat'. The interdependence of indigenous and local communities and biodiversity in local environments means that protecting and encouraging the vitality of biological diversity often requires protecting and encouraging the vitality of local and indigenous communities. Similarly, protecting indigenous cultures and customary practices can help preserve and encourage biological diversity.

Conclusions on Sustainable Use

The world's indigenous and local communities are extremely diverse. It is difficult to make generalizations about these communities, because they do not make up a cohesive group. The lived experience of indigenous and local communities in relation to centralized national governments around the world varies widely. Rather than being united by a common culture or experience, they are united by a common set of aspirations. However, the foregoing discussion does point to some general conclusions about customary practices and how customary sustainable use can be protected and encouraged.

- encouraging and protecting customary sustainable use depends first and foremost on the recognition of the fundamental human rights, including the collective rights of indigenous peoples and local communities;
- recognizing customary use rights can benefit conservation efforts, while denying customary use rights can threaten biological diversity;
- protecting and encouraging customary resource use practices requires a sustainable use approach to conservation;

- acknowledgement of the flux and variability inherent in ecological processes requires an adaptive management approach to sustainable use strategies;
- recognition of the intimate connection between indigenous and local communities and their traditional territories, and recognition of customary tenure systems, is essential to encouraging and promoting customary sustainable resource use; and
- indigenous and local communities must be involved at all levels of resource management, and national governments must be responsive to the input of local communities.

In summary

Traditional knowledge systems, then, are more than compartmentalized sets of information, they are systems of knowledge that exist in spatial, cultural and temporal contexts through which people are able to do such things as interpret reality, make decisions, provide food security, health care and plan for the future. Many bodies of traditional knowledge exist, depending on the particular culture and environment of a people or community, and collectively these bodies of knowledge allow their holders to create new knowledge, to test it and to verify it within the laboratory of daily life and survival. The strength of traditional knowledge is in the test of time.

These verification processes are now becoming more broadly understood, and there is an increasing realization that traditional knowledge systems are viable and legitimate systems of knowledge and should therefore not be cast as subordinate to Western science as a result of the particular cultural space they exist in. Even so, a lack of recognition of the legitimacy of bodies of traditional knowledge and its ongoing lack of adequate legal protection remains an issue, and even provides the basis for a number of the processes that threatens them.

Still, there is now better understanding of the value of traditional knowledge and, in particular, the vital importance that the continued existence of these bodies of knowledge have for the peoples and communities to which they belong. In addition, the value of traditional knowledge in two-way dialogues on global sustainability, not based on the appropriation of traditional knowledge, is gaining greater appreciation.

Of particular importance in a global sustainability dialogue is the realization that not only have indigenous peoples and local communities existed in the most biodiverse environments for a great length of time, but their practices and innovations have been able to be sustained through the application of culturally bound conservation ethics within these environments without undue and deleterious environmental effects for this time. This is a very profound realization, and one that the Western world would do well to gain a greater appreciation of if we are to find our way to a sustainable future for our children and their children after them.

References

Berkes, F. (1999) *Sacred Ecology: Traditional Ecological Knowledge and Management Systems*, Taylor & Francis, Philadelphia and London, p148

Berkes, F., Colding, J. and Folke, C. (2000) 'Rediscovery of traditional ecological knowledge as adaptive management', *Ecological Applications*, vol 10, no 5, pp1251–1262, at p1252

Gadgil, M., Berkes, F. and Folke, C. (1993) 'Indigenous knowledge for biodiversity conservation', *Ambio*, vol 22, nos 2–3, pp151–156, at p151

Hill, R. (2004) *Yalanji Warranga Kaban: Yalanji People of the Rainforest Fire Management Book*, Little Ramsay Press, Cairns

Oviedo, G. and Maffi, L. (2000) *Indigenous and Traditional Peoples of the World and Ecoregion Conservation*, WWF-World Wildlife Fund for Nature, Gland, Switzerland, p1

RAFI (Rural Advancement Foundation International) (1995) *Conserving Indigenous Knowledge: Integrating Two Systems of Innovation*, RAFI and UNDP, New York, p4

Zhang, X. (2004) 'Traditional medicine: Its importance and protection', in S. Twarog and P. Kapoor (eds) *Protecting and Promoting Traditional Knowledge: Systems, National Experiences and International Dimensions*, United Nations Conference on Trade and Development, United Nations Document No. UNCTAD/DITC/TED/10, pp3–6

19

The Global Environment Facility: Our Catalytic Role to Sustain Biodiversity

Monique Barbut

Introduction

The Global Environment Facility (GEF) is the financial mechanism to the most important multilateral environmental agreement that addresses biological diversity: the Convention on Biological Diversity (CBD). This chapter provides an overview of GEF's experience in this role; discusses the current global funding milieu for biodiversity to achieve the objectives of the CBD; and identifies opportunities that the GEF will use to leverage existing funding flows to enhance the generation of global biodiversity benefits.

The CBD provides the global policy framework to address biodiversity issues and also provides the guidance under which the GEF operates to assist developing countries in meeting their obligations under the CBD.

The objectives of the CBD are defined in Article 1 as 'the conservation of biological diversity, the sustainable use of its components and the fair and equitable sharing of the benefits arising out of the utilization of genetic resources, including by appropriate access to genetic resources and by appropriate transfer of relevant technologies, taking into account all rights over those resources and to technologies, and by appropriate funding'.

The GEF funds the additional or 'incremental' cost of generating global biodiversity benefits given that this outcome often comes at a cost above national development and environmental priorities.

The relationship between the Conference of the Parties (COP) of the CBD and the GEF is defined by a memorandum of understanding.[1] In accordance with Article 21 of the CBD, the COP determines policy, strategy, programme priorities and eligibility criteria for access to and utilization of financial resources available through the financial mechanism, including monitoring and evaluation. The GEF defines new or strengthened objectives and approaches, modalities, operational criteria, procedures and any other process needed to meet the objectives of the CBD and to respond to the guidance that the CBD provides. In applying COP guidance in project operations, the GEF and its implementing agencies support country-driven, national priority projects and programmes endorsed by governments.

Resource Requirements and Availability to Achieve the Objectives of the CBD

Estimating the costs of pursuing the three objectives of the CBD has proven to be both difficult and elusive. A notable exception to this are the costs of *in situ* biodiversity conservation, particularly the costs of implementing a globally representative system of protected areas emerging from the collective contribution of national systems of protected areas (PAs). However, cost estimates of achieving the other two CBD objectives do not yet exist.

The tables below give an approximation of CBD-related global expenditure in recent years.

Table 19.1 *Estimates of national expenditures to support CBD objectives (circa 2005, in billions of US dollars)*

Recent national expenditures in biodiversity conservation (mostly PAs)	Recent national expenditures in sustainable use and fair and equitable benefits sharing
1. All developing countries 1.3 to 2.6	All developing countries: unknown
2. All high-income countries 4 to 5	All high-income countries: unknown
3. World total	5.3 to 7.6 World total: unknown

Source: 1. Molnar et al (2004); 2. based on James et al (2001); 3. Pearce (2005, 2007), Gutman and Davidson (2007)

Table 19.2 *Estimates of national and international expenditures in biodiversity conservation (mostly PAs circa 2005, in billions of US dollars)*

A. All developing countries' investments	B. High-income countries' aid to developing countries	C. High-income countries' investments in their own countries	D. Total
1.3 to 2.6	1.2 to 2.5	4 to 5	6.5 to 10

Source: A. Molnar et al (2004); B. Gutman and Davidson (2007); C. based on James et al (2001); D. Pearce (2005, 2007), Gutman and Davidson (2007)

Although no information is readily available that breaks down national funding by sources, international funding, according to sources referenced in Table 19.3 below, may currently range from US$4–5 billion a year. The largest amount of international funds – approximately US$2 billion dollars a year – comes from high-income countries' Overseas Development Assistance (ODA). Most of this is in the form of country-to-country bilateral aid, and the rest is in the form of multilateral aid managed by the GEF, United Nations (UN) agencies, the World Bank and other international financial agencies. The share of ODA going to CBD-related programmes has remained fairly constant through the past 15 years, between 2.5 per cent and 3 per cent of total ODA.

Table19.3 *Estimates of approximate international expenditures to support the three CBD objectives in developing countries, by source (circa 2005, in billions of US dollars)*

1. High-income countries' ODA (both bilateral and multilateral)	2
2. Not-for-profit (including foundations)	~ 1
3. Business and market-based sources	~ 1 to 2
4. Total	~ 4 to 5
5. Of which approximately 30% to 50% went to biodiversity conservation	~1.2 to 2.5
6. And approximately 50% to 70% went to sustainable use and access and benefit sharing	~ 2 to 3.5

Source: 1. from OECD (2007) GEF online database and World Bank. 2. and 3. from Gutman and Davidson (2007); 5. from Table 19.2; 6. difference between 4 and 5

Note: High-income countries' ODA (line 1) is probably an overestimation due to the fact that what the Organisation for Economic Co-operation and Development (OECD) countries report as CBD related-funding may include support to water and sanitation and rural development projects with limited biodiversity value (see OECD, 2007). This, in turn, would result in an overestimation of line 6 (funding for sustainable use and access and benefit sharing) that is obtained by subtracting line 5 from line 4.

Three important conclusions that can be drawn from the financial estimates of support to *in situ* biodiversity conservation:

1. Funding gaps are unevenly distributed among countries and are largest in lower income countries where a significant concentration of the world's biodiversity exists but, due to lower cost structures, absolute funding needs remain lower;
2. Marine systems are likely the most underfunded ecosystems, in terms of both current management and expansion of PAs; and
3. Estimates of the cost of expansion to create a globally representative PA system are clearly imprecise, however, sufficient information exists to

estimate the resources required for the creation of well-managed, comprehensive PA system that will bring us closer to the globally agreed targets.

Resources Needed for a Global Network of PAs

Recent analysis has produced a range of estimates of funding needs and gaps for the establishment and management of a global network of PAs. Low-end estimates of the required resources for establishing and managing an expanded PA system in tropical countries range from US$1.1 billion per year (Vreugdenhil, 2003) to US$4 billion per year (Bruner et al, 2004). Medium-range estimates (James et al, 1999, 2001) propose that the cost of management of the world's existing PAs would be US$14 billion (including management costs and opportunity costs) and US$20–25 billion a year for an expanded representative system. Balmford et al (2004) estimate that a global PA network encompassing 15 per cent of the world's total land area and 30 per cent of its ocean area would cost some US$45 billion a year over 30 years, including management costs and opportunity costs.

In summary, if we take as a given that the current estimated annual expenditure globally for biodiversity conservation (US$6.5–10 billion, see Table 19.2) is mainly directed towards PA management and assume the conservative mid-range estimate proposed above for managing the existing protected areas, global investment is between 50–70 per cent of what is required to manage existing protected areas. This gap would be larger when considering the 2010 target for terrestrial protected areas, 'a global network of comprehensive, representative and effectively managed national and regional protected area system is established' (CBD, 2000) and the 2012 target for marine areas, 'a global network of comprehensive, representative and effectively managed national and regional protected area system is established' (CBD, 2000). Given that funding gaps are largest in lower income countries, where a significant concentration of the world's biodiversity exists, closing these gaps in particular is bound to bring us closer to the 2010 and 2012 CBD targets.

Traditional and New Sources of Financing for Biodiversity

High-income countries pay for all and middle-income countries pay for most of their CBD-related investments. For low-income countries, international sources of funding are vital. Public budgets are still the largest source of funds, but other sources are increasingly important. For example, at the national level in some developing countries, conservation concessions, payment for ecosystem service (PES) schemes, and conservation easements are being applied, and have the potential to contribute substantially more in the future.

Not-for-profit funding, coming from international conservation non-governmental organizations (NGOs), private foundations and business-related foundations, may contribute more than US$1–1.5 billion annually to international investment in biodiversity conservation, although precise figures are difficult to identify. As in the case of ODA, not-for-profit sources for biodiversity conservation have grown sluggishly during the past decade, which may be a reflection of limited awareness and interest in biodiversity as well as the state of the economy and a reduction in available funds for philanthropy in general. Furthermore, the impacts of climate change and the need to mitigate or adapt to these changes has recently captured global interest and funding has followed.

The two major sources of international market-based funding for biodiversity conservation and sustainable use are: (a) international visitors, ecotourism and tourism; and (b) markets for environment-friendly products – organic and/or certified agricultural products, timber, coffee, fish, and ecotourism – through a range of certification systems such as the International Federation of Organic Agriculture Movements (IFOAM), the Forest Stewardship Council, Rainforest Alliance, and the Marine Stewardship Council (MSC).

The incipient field of international payments for ecosystem services (e.g. bio-prospecting and biocarbon) has triggered high expectations, but thus far has produced limited funding for biodiversity programmes. However, the progress being made at the level of the United Nations Framework Convention on Climate Change (UNFCCC) with schemes related to the proposed Reduced Emissions from Deforestation and Forest Degradation Plus (REDD+) approaches and associated financial mechanisms that seek multiple benefits in the areas of carbon dioxide emissions mitigation, biodiversity conservation and enhanced rural livelihoods, are all poised to attract significant resources during the next several years.

While the sources of funding for CBD-related investments are few – governments, NGOs, private philanthropic foundations, businesses and households – the financial mechanisms involved could be in the hundreds. Table 19.4 presents an aggregated picture of both traditional and more innovative financial mechanisms available for biodiversity funding at local, national and international levels, many of which the GEF is actively promoting through its biodiversity programme.

As we consider the future of biodiversity funding, four future trends are suggested:

1. Among the traditional financial mechanisms at the local, regional, national and international level: government budgets, tourism revenues and NGO programmes are important and have opportunities to grow and innovate in the future.
2. Some of the more promising innovative mechanisms at local, national and regional level are: promoting payments for ecosystem services, particularly in the area of carbon credits for avoided deforestation and reforestation, markets for green products, new forms of charity, and businesses engagement in biodiversity conservation (for example, private/public partnerships, philanthropic ventures with the private sector, biodiversity offsets, and so on).

Table 19.4 *A summary of traditional and innovative financial mechanisms*

More traditional	More innovative
Local level financial mechanisms	
• Local or state government budget allocations • Earmarking public revenues • *Protected areas entrance and fees* • *Tourism-related incomes* • Local NGOs and charities • Local businesses goodwill investments	• *Local green markets* • *Local markets for modalities of ecosystem services[2] (PES)* • Tax systems at municipal and state levels
National (and Regional) Level Financial Mechanisms	
• Government budgetary allocations • Earmarking public revenues • *National tourism* • *National NGO grant-making* • National businesses goodwill investments • National private foundations • *Environmental funds*	• Environmental tax reform • Reforming rural production subsidies • *National level PES* • Green lotteries • New goodwill fund-raising instruments (internet-based, 'round-ups',[3] and so on) • *Businesses/public/NGO partnerships* • *Businesses voluntary standards* • *National green markets* • *National markets for all types of ecosystem services (PES)* • Funding marketplaces to bring quality projects and programmes to interested donors
International (and regional) level financial mechanisms	
• Bilateral aid • Multilateral aid • Debt-for-Nature Swaps • Development banks and agencies • GEF • UN Agencies • International NGO grant-making • Private foundations • International tourism • Businesses goodwill investments • *Environmental funds*	• Long-term ODA commitments • Auction or sale of part of the carbon emission permits and other cap-and-trade schemes • Environment-related taxes • Sharing in international solidarity taxes • Reforms of the international monetary system • Green lotteries • New goodwill fund-raising instruments (internet-based, 'round-ups', and so on) • *Businesses/public/NGO partnerships* • *Businesses voluntary standards* • *International green markets* • *International markets for certain modalities of ecosystem services (PES)* • Dedicated public funds • Funding marketplaces to bring quality projects and programmes to interested donors • *GEF's new Sustainable Forest Management Programme that combines funds from different GEF focal areas (biodiversity, climate change, land degradation) to generate global biodiversity benefits*

Note: Mechanisms currently being strengthened through GEF projects are in italics.

3. Among innovative financial mechanisms at the national level, environmental tax reforms and the reform of production subsidies can become important sources of funding for biodiversity investments in high and medium-income countries; but less so in low-income countries where the tax base is usually small.
4. Among innovative financial mechanisms at the international level several donor countries and other stakeholders are already experimenting with, or discussing long-term ODA commitments, environmental taxes, international solidarity taxes, green lotteries, promoting payments for ecosystem services and markets for green products.

GEF's Investment Strategy to Enhance Leverage for Biodiversity Conservation and Sustainable Use

Despite this challenging financial backdrop, the GEF has remained committed to advancing the global biodiversity agenda in its recipient countries. The GEF's biodiversity portfolio has been the largest project portfolio in terms of grant money awarded since the GEF began operations. Since 1991, the GEF has provided about US$2.9 billion in grants and leveraged an additional US$8.2 billion in co-financing in support of 990 projects that addressed the loss of globally significant biodiversity in more than 155 countries. In the recently completed GEF-4 (2006–2010), GEF investment in biodiversity totalled US$895 million, which leveraged an additional US$2.4 billion.

The GEF is the largest funding mechanism for PAs worldwide and has invested in more than 2302 protected areas, covering more than 634 million hectares. The GEF has provided more than US$1.89 billion to fund protected areas management, leveraging an additional US$5.95 billion in co-financing from project partners for a total of almost US$8 billion. More than 70 countries have benefited from GEF investments to enhance financial sustainability of protected area systems as GEF has increased its emphasis on systemic approaches to strengthening PAs. The GEF is recognized as a pioneer in this area and has supported more than 40 conservation trust funds worldwide, investing more than US$300 million in total. In addition, the GEF has supported more than 30 payments for ecosystem services schemes to provide steady, reliable funding for PA management and biodiversity conservation.

The GEF realizes that civil society plays a crucial role in advancing biodiversity conservation and has therefore developed strong partnerships with civil society organizations, including NGOs and indigenous and local communities, through its biodiversity programme. The GEF Small Grants Programme has provided 7042 small grants to NGOs and community-based organizations in 121 countries, with a total GEF funding of US$153 million, which has leveraged an additional US$219 million. In addition to advancing biodiversity conservation and sustainable use, many of these community-driven projects make key contributions to reducing poverty in rural communities.

The Critical Ecosystem Partnership Fund (CEPF), another GEF partnership mechanism, has reached out to more than 1550 civil society organizations in 51 countries to help conserve 18 of the world's most important biodiversity hot spots. CEPF has invested US$120 million to date that has leveraged an additional US$261 million. Leveraging other resources through the vehicle of GEF grants will always remain a critical aspect of enhancing GEF's impact and making GEF an effective resource mobilization mechanism in its own right. Part of GEF's role as the largest funder for global biodiversity conservation is to leverage impact through targeted responses to biodiversity loss. This strategic approach to investment is embedded in GEF's biodiversity strategy, which incorporates almost two decades of operational experience, programme evaluation findings and the latest advances in conservation practice. The strategy incorporates three key principles to achieve lasting biodiversity conservation and sustainable use:

1. sustainability of results and the potential for replication are emphasized;
2. strategic approaches that strengthen country-enabling environments (policy and regulatory frameworks, sustainable financing, institutional capacity-building, science and information, awareness) are central components of project interventions; and
3. biodiversity conservation and sustainable use are mainstreamed in the wider economic development context.

The Millennium Ecosystem Assessment (MEA) identified the most important direct drivers of biodiversity loss and degradation of ecosystem goods and services as habitat change, climate change, invasive alien species (IAS), overexploitation, and pollution. These drivers are influenced by a series of indirect factors of change including demographics, global economic trends, governance, institutions and legal frameworks, science and technology and cultural and religious values. The GEF biodiversity strategy addresses a subset of the direct and indirect drivers of biodiversity loss and focuses on the highest leverage opportunities for the GEF to contribute to sustainable biodiversity conservation.

The goal of the biodiversity strategy is the conservation and sustainable use of biodiversity and the maintenance of ecosystem goods and services. To achieve this goal, the strategy encompasses five objectives:

1. improve the sustainability of protected area systems;
2. mainstream biodiversity conservation and sustainable use into production landscapes/seascapes and sectors;
3. build capacity to implement the Cartagena Protocol on Biosafety;
4. build capacity on access to genetic resources and benefit sharing; and
5. integrate CBD obligations into national planning processes through enabling activities.

Leveraging Sustainable Forest Management (SFM)/REDD+ to Generate Global Biodiversity Benefits

Forest ecosystems provide a variety of benefits which are realized at the global, subregional, national and local scales. Beyond their key role in climate change mitigation of land-based emissions, forests harbour a significant fraction of the world's biodiversity wealth and are responsible for the provision of key ecosystem services, including functioning as carbon sinks and storehouses, as buffers against soil degradation and desertification, as well as sustaining the livelihoods of hundreds of millions of rural people everywhere. These linkages imply that forests can be conserved and managed for multiple benefits.

The conversion and degradation of tropical forests, which accounts for approximately 90 per cent of the total greenhouse gas (GHG) emissions from deforestation and for nearly 80 per cent of the threats to biodiversity globally, has been made the focus of an innovative experiment conducted as part of GEF's SFM programme. Through this initiative, countries were provided incentives to invest in SFM and Land-use, Land-use Change and Forestry (LULUCF) activities. The incentive mechanism was resourced by reserving portions of the funding windows of biodiversity and climate change, complemented by land degradation resources, and directed to SFM activities.

The investment strategy in SFM for GEF-5 will build on the successful experience with the SFM portfolio development gained in GEF-4. In GEF-5, the expanded SFM/REDD+ programme will be established as an incentive mechanism for all forest countries to invest in all forest types using resources from the GEF focal areas of biodiversity, climate change, land degradation and, when appropriate, from international waters (transboundary watersheds) towards transformative programmatic investments. A US$250 million funding envelope for SFM, based on the GEF-4 experience, would result in a total GEF investment in SFM/REDD+ for GEF-5 approaching US$1 billion by the end of the next funding cycle, before co-financing is considered, which could leverage this amount threefold.

GEF-5 Replenishment: Good News Going Forward

In May 2010, the GEF received its largest ever increase in funding, with more than 30 nations pledging US$4.28 billion for the next four years. This represents a 54 per cent increase in new resources provided by donors.

In the case of biodiversity, funding pledged increased from US$941 million in GEF-4 to US$1.21 billion in GEF-5, an increase of 29 per cent. In addition, a revised system for resource allocation (System for Transparent Allocation of Resources, STAR) in GEF-5 provides for more resources for marine ecosystems, which historically have been underfunded.

This robust replenishment has maintained the GEF's position as the largest donor advancing the cause of biodiversity globally. Leveraging these resources will remain a focus of the GEF through creative investing that generates a maximum of global biodiversity benefits and by continuing to identify and support innovations in conservation finance that can be replicated worldwide. This effort of the GEF[4] will be another important chapter of the steady and ongoing contribution we have made to help the global community realize the aspirations embedded in the objectives of the CBD.

Notes

1 Decision III/6: Memorandum of Understanding between the Conference of the Parties to the Convention on Biological Diversity and the Council of the Global Environment Facility.
2 Markets for environment-friendly products, organic and/or certified agricultural products, timber, coffee, fish and ecotourism.
3 'Round-ups' are schemes where households, employees or buyers agree that their bills be rounded up to the next dollar and that the difference between their purchase or expenditure and the round-up be donated to charity.
4 The GEF unites 181 member governments – in partnership with international institutions, NGOs and the private sector – to address global environmental issues. An independent financial organization, the GEF provides grants to developing countries and countries with economies in transition (EIT) for projects related to biodiversity, climate change, international waters, land degradation, the ozone layer and persistent organic pollutants. These projects benefit the global environment, linking local, national and global environmental challenges, and promoting sustainable livelihoods. Established in 1991, the GEF is today the largest funder of projects to improve the global environment. The GEF has allocated US$9.2 billion, supplemented by more than US$40 billion in co-financing, for more than 2600 projects in more than 165 developing countries and countries with economies in transition. Through its Small Grants Programme, GEF has also made more than 10,000 small grants directly to NGOs and community organizations. The GEF partnership includes ten agencies: the United Nations Development Programme (UNDP); the United Nations Environment Programme (UNEP); the World Bank; the Food and Agriculture Organization (FAO); the United Nations Industrial Development Organization; the African Development Bank; the Asian Development Bank; the European Bank for Reconstruction and Development; the Inter-American Development Bank (IDB); and the International Fund for Agricultural Development (IFAD). The Scientific and Technical Advisory Panel provides technical and scientific advice on GEF's policies and projects.

References

Balmford, A., Gravestock, P., Hockley N., McClean, C. J. and Roberts, C. M. (2004) 'The worldwide costs of marine protected areas', *Proceedings of the National Academy of Sciences*, vol 101, no 26, pp9694–9697

Bruner, A. G., Gullison, R. E. and Balmford, A. W. (2004) 'Financial costs and shortfalls of managing and expanding protected areas systems in developing countries', *BioScience*, vol 54, no 12, pp1119–1126

CBD (Convention on Biological Diversity) (2000) www.cbd.int/protected/pow/learnmore/goal11

Gutman, P. and Davidson S. (2007) *A Review of Innovative International Financial Mechanisms for Biodiversity Conservation With a Special Focus on the International Financing of Developing Countries' Protected Areas*, WWF-MPO, Washington, DC

James, A. N., Gaston, K. J. and Balmford, A. (1999) 'Balancing the Earth's accounts', *Nature*, vol 401, no 6751, pp323–324

James, A. N., Gaston, K. J. and Balmford, A. (2001) 'Can we afford to conserve biodiversity?', *BioScience*, vol 51, no 1, pp43–52

Molnar A., Scherr, S. J. and Khare, A. (2004) *Who Conserves the World Forests? Community Drive Strategies to Protect Forests and Respect Rights*, Forest Trends, Washington, DC

MNP/OECD (2007) 'Background report to the OECD Environmental Outlook to 2030. Overviews, details and methodology of model-based analysis', Netherlands Environmental Assessment Agency Bilthoven, The Netherlands and Organisation of Economic Co-operation and Development, Paris, France

Pearce, D. W. (2005) 'Paradoxes in biodiversity conservation', *World Economics,* vol 6, no 3, pp57–69

Pearce, D. W. (2007) 'Do we really care about biodiversity?', *Environmental Resource Economics*, vol 37, no 1, pp313–333

Vreugdenhil, D. (2003) 'Modeling the financial needs of protected area systems: An application of the "minimum conservation system" design tool', paper prepared for the 5th World Parks Congress, Durban, South Africa, 8–17 September

20

Is 'Biodiversity' the Next 'Climate Change' for Business?

Craig Bennett

Introduction

Think 'nature conservation' and most people will think of what is known within the professional conservation community, somewhat mischievously, as the 'big charismatic mega-fauna'.

There is little doubt that it has been a relatively small group of large, big eyed, often fluffy animals that has grabbed most of the attention (and money) for conservation down the years. There is also little doubt that the established narrative for protecting bears, elephants, tigers, rhinos, whales, dolphins and nature more generally has been, to date, one largely based on a moral argument rather than an economic one. Perhaps we should not be so surprised, then, that the protection of ecosystems and biodiversity has been an issue that most of the business community have traditionally seen as irrelevant to the good old-fashioned business of business. But as the understanding of the *role* that biodiversity and ecosystems play in supporting our economy grows, so some parts of the business community are starting to look at the issue with fresh eyes.

Over the past couple of decades, climate change has moved from an issue that never made it into the corporate boardroom to one that, directly or indirectly, is now influencing many decisions of corporate strategy. Could it be that 'biodiversity' is now set to follow a similar issue trajectory? Is 'biodiversity' the next 'climate change' for business?

'Dangerous' Climate Change versus 'Beneficial' Biodiversity

The global challenge of climate change has received unprecedented levels of political and corporate attention over the past decade, but the parallel issue of biodiversity loss has been largely ignored (at least by comparison). There are several explanations for this. The first is the very different contextual framing of these two global challenges, which is evident in the contrasting objectives for the United Nations (UN) conventions established to address either issue. The objective of the UN Framework Convention on Climate Change (UNFCCC) (as articulated in Article 2) states that:

> *The ultimate objective of this Convention and any related legal instruments that the Conference of the Parties may adopt is to achieve, in accordance with the relevant provisions of the Convention, stabilization of greenhouse gas concentrations in the atmosphere at a level that would prevent dangerous anthropogenic interference with the climate system. Such a level should be achieved within a time frame sufficient to allow ecosystems to adapt naturally to climate change, to ensure that food production is not threatened and to enable economic development to proceed in a sustainable manner.* (UN, 1992a)

It is a purpose born out of fear, whereas the objective of the UN Convention on Biological Diversity (CBD) (as articulated in Article 1) is one of complicated expectation:

> *The objectives of this Convention, to be pursued in accordance with its relevant provisions, are the conservation of biological diversity, the sustainable use of its components and the fair and equitable sharing of the benefits arising out of the utilization of genetic resources, including by appropriate access to genetic resources and by appropriate transfer of relevant technologies, taking into account all rights over those resources and to technologies, and by appropriate funding.* (UN, 1992b)

The contrast in the contextual framing of these two objectives is stark and reflects much of the political discourse concerning both issues to date. 'Climate change' is quite rightly presented as an urgent and 'dangerous' threat to humanity with a (relatively) straightforward solution; the stabilization of greenhouse gas (GHG) concentrations. The failure to achieve this within a sufficient time frame will, the UNFCCC implies, threaten food production and economic development; something of concern to rich and poor countries alike. In contrast, the CBD is presented as an end in its own right; one that delivers undefined 'benefits' that must be shared 'equitably'. What the CBD's objective neglects to do is set out the risks to humanity of failing to conserve ecosystems.

While this more positive approach might be appealing and appropriate from an ideological perspective, it is easy to see how such a different context for the issue has placed it lower on the political agenda. If we stop to think about what normally brings governments together in common practical purpose, it is usually to address urgent threats such as war, terrorism, recession or disease epidemics rather than to realize undefined 'benefits'.

Furthermore, because many of the world's biodiversity 'hot spots' are to be found in poorer countries, there is probably an implicit (and false) political assumption that most of the benefits of maintaining biodiversity are irrelevant to industrialized nations.

Climate change has not secured as much action from governments as it deserves, but that which it has secured has been because it is correctly perceived as a threat to humanity. But biodiversity has received even less attention because, for too long, it has merely been seen as a 'nice to have' (for the poorer countries) rather than 'dangerous to lose' (for all nations).

This is demonstrated by the rising concern regarding tropical deforestation, arguably the one area of biodiversity loss that has received some sort of international political traction in recent years (such as in numerous G8 communiqués). This has happened because of the growing realization of the very significant contribution that continued deforestation makes to climate change (around 17 per cent of annual global GHG emissions) (IPCC, 2007) and how it is therefore of concern to all countries, not just those with tropical forests.

If biodiversity is to follow the same issue trajectory as climate change, it will be important to build the same sort of scientific and political consensus about the risks associated with ecosystem degradation and loss and, crucially, how these represent risks that should be of concern for poor and rich countries alike.

From 'Species' to 'Biodiversity' to 'Ecosystem Services'

Much of the modern conservation movement grew out of the popularity for natural history that blossomed in the 19th century. Like many of the equivalent museums around the world, the Natural History Museum in London contains tens of thousands of perfectly preserved butterflies, beetles and other insects that amateur naturalists collected from distant lands as they tried to catalogue and make sense of the natural world throughout the 1800s.

For well over a century, the focus of conservationists was firmly on species rather than ecosystems, let alone ecosystem services. The International Union for Conservation of Nature (IUCN) Red List of Threatened Species, for example, is recognized as the most comprehensive, objective global approach for evaluating the conservation status of plant and animal species. Its goals are to identify and document those species most in need of conservation attention if global extinction rates are to be reduced, and to provide a global index of the state of change of biodiversity (IUCN, 2010).

Although species are often chosen because they represent good 'biodiversity indicators' for the health of their ecosystems, the human propensity to focus on quantitative measures means that it is often the cumulative number of species in each category (and critically the number of species extinctions) that generates most attention.

Rockström et al (2009) focused on extinction rates in their recent attempt to identify and quantify 'a safe operating space for humanity', although they note its limitations:

> *The fossil record shows that the background extinction rate for marine life*

is 0.1–1 extinctions per million species per year; for mammals it is 0.2–0.5 extinctions per million species per years. Today, the extinction rate of species is estimated to be 100 to 1000 times more than what could be considered natural. As with climate change, human activities are the main cause of the acceleration...

There is growing understanding of the importance of functional biodiversity in preventing ecosystems from tipping into undesired states when they are disturbed. This means that apparent redundancy is required to maintain an ecosystem's resilience...

Although it is now accepted that a rich mix of species underpins the resilience of ecosystems, little is known quantitatively about how much and what kinds of biodiversity can be lost before that resilience is eroded... Ideally, a planetary boundary should capture the role of biodiversity in regulating the resilience of systems on Earth. Because science cannot yet provide such information at an aggregate level, we propose extinction rate as an alternative (but weaker) indicator. As a result, our suggested planetary boundary for biodiversity of ten times the background rates of extinction is only a very preliminary estimate. More research is required to pin down this boundary with greater certainty. However, we can say with some confidence that Earth cannot sustain the current rate of loss without significant erosion of ecosystem resilience. (Rockström et al, 2009)

Individual species are important. Each of those big charismatic mega-fauna that have historically played such an important role on conservation fund-raising leaflets, also play a critical role within their respective ecosystems. As top predators, tigers (*Panthera tigris*) and killer whales (orca) (*Orcinus orca*) help regulate the populations of their prey, whether this be sambar deer (*Rusa unicolor*) in an Indian jungle or Humbolt squid (*Dosidicus gigas*) off the Pacific coast of California. European brown bears (*Ursus arctos*) play an important role in seed dispersal of some species of fruiting tree such as plum trees because the plum stone is too large for most other mammals to swallow.

But, despite the historic focus of the conservation community on species, in most cases it is the ecosystems that they form an integral part of that are really important for humanity. In the same way that individual species have a role to play within their ecosystems, whole ecosystems have an important role within the larger planetary system.

The Millennium Ecosystem Assessment (MEA), published in 2005, represented a major step forward in presenting an emerging scientific consensus about the risks associated with ecosystem degradation and loss because of its focus on 'ecosystem services'. The MEA offered the following definitions for 'ecosystem' and 'ecosystem services':

An ecosystem is a dynamic complex of plants, animals, microbes, and physical environmental features that interact with one another. Ecosystem

> *services are the benefits that humans obtain from ecosystems, and they are produced by the interactions within the ecosystem.* (MEA, 2005a)

The MA noted that ecosystems such as forests, grasslands, mangroves and urban areas provide different services to society including:

- provisioning services (such as food, water and wood);
- regulating services (such as climate, flood and disease regulation);
- cultural services (such as aesthetic, recreational and educational); and
- supporting services (needed to maintain other services, such as nutrient cycling and soil formation).

Some of these ecosystem services are more obvious (food and water) than others (disease regulation). Some are local (provision of pollinators), others are regional (flood control or water purification) while others are global (such as climate regulation). All of them are critical for the economy, for business, for humanity. In every case, however, the economic 'value' of these services has rarely been taken into account in economic and business decision-making.

The MEA represented a first step in raising the profile of ecosystem services, in a way that can perhaps be compared to the role that the Intergovernmental Panel on Climate Change's (IPCC) First Assessment Report played for climate change when it was published in 1990.

The MEA was not 'owned' or approved by the world's governments in quite the same way that IPCC reports have been, however, and so did not gain the same level of political purchase. Current plans to establish an Intergovernmental Science-Policy Platform on Biodiversity and Ecosystem Services (IPBES), already welcomed by the G8 (G8, 2010), could come to represent the stronger, ongoing, international science–policy platform that is now urgently needed to enable emerging scientific knowledge to be translated into specific policy action at the appropriate levels (IPBES, 2010). A critical first test for IPBES will be whether it is able to raise the political importance of biodiversity ahead of the Rio+20 Earth Summit in May 2012.

The Economics of Climate Change and Biodiversity Loss

It was the growing recognition of climate change as an economic rather than just an environmental issue that did most to raise it up the political and corporate agenda, however. Although this came about because of numerous contributions (not least the reports produced by Working Group 2 of the IPCC), the most notable was *The Stern Review on the Economics of Climate Change*, a 700-page report released for the UK government on 30 October 2006 by the then head of the UK Government Economic Service and a former chief economist of the World Bank, Nicholas Stern.

To my mind, there were two aspects about *The Stern Review* that enabled it to secure political and corporate traction in a way that other initiatives failed to do. The first is to do with the substance of the report. Its overall conclusions were clear and simple, stark but hopeful: 'The benefits of strong and early action far outweigh the economic costs of not acting.' Even the conclusions from the more formal economic models used in *The Stern Review* could be easily understood by decision-makers and had a global relevance:

> *The Review estimates that if we don't act, the overall costs and risks of climate change will be equivalent to losing at least 5% of global GDP each year, now and forever. If a wider range of risks and impacts is taken into account, the estimates of damage could rise to 20% of GDP or more. In contrast, the costs of action – reducing greenhouse gas emissions to avoid the worst impacts of climate change – can be limited to around 1% of global GDP each year.* (Stern, 2006, pXV)

The second aspect that enabled *The Stern Review* to secure political and corporate traction has nothing to do with the substance of the report, but everything to do with its political context. *The Stern Review* was commission by the UK government's finance ministry (Her Majesty's Treasury) and *not* by the environment ministry. When it was launched, it was championed by the then Prime Minister, Tony Blair, and finance minister, Gordon Brown, *not* merely the environment minister:

> *Nicholas Stern's review will be seen as a landmark in the struggle against climate change. It gives a stark warning, but also offers hope. It proves comprehensively that tackling climate change is a pro-growth strategy. The economic benefits of strong early action easily outweigh the costs. The framework for action laid out by the Review is both ambitious and realistic. Climate change is a global problem and Stern's conclusions are a wake-up call not just for the UK, but for every country in the world.* (Tony Blair, in Stern, 2007, pii)

> *The Stern Review on the Economics of Climate Change is the most comprehensive analysis yet, not only of the challenges, but also of the opportunities from climate change. Stern makes clear that climate change is a global challenge that demands a global solution. Above all, environmental policy is economic policy. It is my hope that this Review is discussed and understood as widely as possible, throughout the world, not just by Governments, but also by business leaders, NGOs, international institutions and society as a whole.* (Gordon Brown, in Stern, 2006, pii)

If biodiversity is to follow the same issue trajectory as climate change, its advocates must now carve out similarly clear and simple messages, and secure the highest levels of mainstream political support.

The Economics of Ecosystems and Biodiversity (TEEB) study, led by the leading economist and banker Pavan Sukhdev, represents an important step forward in this regard because it attempts to calculate the economic value of ecosystem services.

Sukhdev's team has systematically calculated the economic significance of conserving hundreds of different ecosystem types around the world. Muthurajawela Marsh, for example, is a coastal wetland in a densely populated area of Sri Lanka. The TEEB study has calculated that the 'provisioning services' provided by the marsh (such as agriculture, fishing and fuel wood) raise a local income of around US$150 per hectare per year) but that this is dwarfed by the ecosystem's value in preventing flooding over a much wider area, which is worth around US$1900 per hectare per year, while its role in treating industrial and domestic wastewater is worth around US$650 (TEEB, 2010).

In another study for TEEB, the value of the ecosystem services provided by coral reefs is estimated at around US$189,000 per hectare per year for storm protection, around US$1 million for tourism, around US$57,000 for genetic materials and bio-prospecting and up to US$3800 for fisheries (TEEB, 2010).

The Stern Review estimated that the cost of tackling climate change would be in the region of 1–2 per cent of annual global wealth, but that the longer term benefits would be 5–20 times that figure. The indications are that Sukdev's team will conclude that the costs of protecting the planet's ecosystems will be higher, but so will the benefits; between 10 and 100 times the cost of protection (Jowit, 2010).

One key challenge for TEEB will be to ensure their findings are picked up and owned by the most senior of mainstream political leaders, not just environment ministers. Unlike *The Stern Review*, which was proposed and commissioned by a finance ministry, TEEB was proposed by G8+5 environment ministers and only then endorsed by G8+5 leaders at the Heiligendamm Summit in 2007. It has been hosted by the United Nations Environment Programme (UNEP) and funded by the environment and/or development ministries of the European Commission, Germany, the UK, Norway, The Netherlands and Sweden (not the finance ministries). Although the collaborative and international aspects of TEEB are welcome, is there a danger that other, more powerful ministries within these governments will dismiss it as yet another environmental report?

Another key challenge for TEEB will be to demonstrate that the economic costs of ecosystem degradation are important to industrialized economies not just poorer nations, and to specific economic actors (such as corporations) not just society as a whole. Or rather, that these costs are important enough to industrialized countries that it will seem inevitable that nations will have to agree to introduce the policy frameworks needed to correct this particular market failure and so make the costs of ecosystem degradation relevant to every economic actor.

Mapping the Landscape of Corporate Risk and Opportunity

To do this, it is important to understand how the landscape of corporate risk and opportunity has evolved on climate change, and what this might mean for biodiversity and ecosystem services.

From 2007 to 2010, I was director of The Prince of Wales's Corporate Leaders Group (CLG) on Climate Change, an initiative of the University of Cambridge Programme for Sustainability Leadership (CPSL). The CLG brings together business leaders who believe that there is an urgent need to develop new and longer term policies for tackling climate change. The initiative is cross sector, encompassing energy producers, manufacturers, banks, retailers, utilities and others. It has been active since 2005 in progressing action on climate change, working with national governments, international fora and within the business community (CPSL, 2010).

The standard mantra is that companies do not like or wish for regulation. But during the almost four years in which I was running the CLG – a cross sector business initiative proactively seeking regulation – I came to recognize five categories of business case in support of regulation to tackle climate change:

1. **Policy clarity in the medium to long term:** This is most relevant to those companies that are heavy emitters (such as energy companies, cement manufacturers or airlines) or companies that find themselves in sectors with high-carbon intensity (such as airport operators). For these companies, climate change represents one of the biggest strategic challenges they face. Once business leaders in these companies have accepted that government action to address climate change (and their own company's emissions) is inevitable, they want clarity as to how and when that change will happen (for example, targets and timescales) and they want the opportunity to shape the policy response and adjust corporate strategy and capital investments accordingly.
2. **New markets for low-carbon technologies:** For many companies, there are significant business opportunities to be realized by moving to a low-carbon economy (for example, manufacturers of fuel cells or energy efficient lighting, renewable energy companies or low-carbon transport solutions). These companies want governments to introduce the policy frameworks that will create a new market for low-carbon goods and services, provide them with the confidence to scale-up investments, stimulate innovation and regulate out high-carbon alternatives.
3. **Protecting or enhancing corporate image, reputation and brand:** Companies with strong brand recognition, large numbers of consumers and employees (for example, retailers and consumer product manufacturers) will want to be seen to be playing a proactive, leadership role in climate change debate. These companies often find themselves on the 'consumer front line' and will want to avoid getting blamed for society's failure to address sustainability concerns. They will also crave consumer and supply chain security. Some

will want policy frameworks that help them 'choice-edit' out those products that work against this.

4. **Seeking long-term economic security and managing risk**: Banks, pension funds and insurance companies became concerned about climate change because of Nick Stern's predictions regarding how, if left unchecked, it could result in loss of 20 per cent of global gross domestic product (GDP). These companies want governments to respond with policy frameworks that protect the long-term value of their investments, support steady 'sustainable economic growth' and help them minimize (or at least manage) risk in the longer term.
5. **Adaption and increased resilience of corporate operations and infrastructure**: Many companies providing an essential service (for example, water and energy companies, food retailers and mobile phone operators) want to see climate adaptation policy frameworks that will enable society to adapt to future changes in the climate, increase resilience to extreme weather events, and improve management of scarce resources (for example, water). Some of these companies have already experienced the strain that extreme weather events can put on their ability to deliver their goods or services to market; many more are concerned that their business models are threatened unless society becomes more resilient to a changing climate.

It may be that all five of these categories are relevant for some companies, for others it will be fewer. But it is unlikely that at least one of them is not relevant for a major company in one way or another.

It is now imperative that business develops an understanding of the categories of business risk and opportunity that are pertinent to the loss of biodiversity and ecosystem services. The MEA (2005b) has offered six headings of 'How the MEA findings affect your bottom line' (summaries by the author):

1. **Licence to operate**: A broad range of stakeholders including local communities, regulators, investors, employees and society at large have increasing expectations that companies will manage their impacts on ecosystems in a sophisticated manner or, as TEEB (2010) has suggested, to deliver 'net positive impact'. Failure to meet these expectations in a proactive way and to involve stakeholders in decision-making may result in companies being denied access to natural resources, either through regulatory or other means. This category of risk is particularly important to agricultural or mineral extractive companies.
2. **Corporate image, reputation and brand risk**: Much the same drivers as the equivalent category of risk for climate change but biodiversity provides an added level of risk because of the media interest in charismatic species and habitats (for example, the brand damage that some food companies have suffered after being associated with declining orang-utan populations because of the use of palm oil in their products).

3. **Cost of capital and perceived investor risk:** Investors will steer away from those sectors and firms whose risks and potential liabilities relating to biodiversity and ecosystem services are not well understood.
4. **Access to raw materials:** As pressure on ecosystem services grows, businesses may find themselves increasingly competing within and between industry sectors and with communities for dwindling resources (for example, water, agricultural land, timber). Retailers are exposed to this category of risk through supply chains and are vulnerable to sudden commodity price fluctuations or supply shortfalls.
5. **Operational impacts and efficiencies:** As ecosystem services become more regulated, companies will need to become more efficient in their use of land, energy and water resources. Commercial advantages will go to those companies that are more efficient than their competitors.
6. **New business opportunities:** New categories of product will emerge as consumers express preferences based on their desire to conserve biodiversity, such as organic farming and ecotourism. Government incentive programmes and regulatory mechanisms will create markets to support payments for ecosystem services.

Much more work is needed to better understand the categories of corporate risk and opportunity relating to biodiversity and ecosystems services, including the ways they are similar and different to those that have been articulated in relation to climate change. But as business leaders come to experience and appreciate these risks and opportunities, so the corporate response will unfold.

The Corporate Response

As noted in a previous paper (Bennett, 2009), the sustainability challenge requires nothing short of transformational change to the way we manage our economy. Incremental change (that which can be delivered by companies acting alone) will fail to address the scale and urgency of problems like climate change, breakdown of ecosystem services, water scarcity and growing inequality.

Transformational change will only be possible through government interventions that are guided by a clear, consistent and strategic approach to policy-making – and where long-term regulatory frameworks are put in place to correct market failure and to change the terms of trade for resource consumption. Governments will only be able to introduce such bold and long-term policy measures if they are given support and political space by the business community to do so, and so the role that companies can play as progressive 'corporate citizens' – urging governments to introduce transformational policies – becomes critical.

The corporate sector's history in public policy has not been good. The public image of a 'corporate lobbyists' is of someone working to protect a narrow vested interest, often against the interests of society as a whole. There are plenty of famous examples of companies, or groups of companies spending millions of dollars in opposition to progressive public policy, whether it is tobacco companies in the 1960s and 1970s questioning the link between smoking and cancer, aerosol companies lobbying against the banning of chlorofluorocarbons (CFCs) in the 1980s, or oil companies clubbing together as the Global Climate Coalition to question the science of climate change in the 1990s. There has long been a widespread perception of 'corporate influence' as a malign force (as described in Monbiot, 2000, and Korten, 1996).

Over the past decade, however, the business community has demonstrated that it includes a good proportion of business leaders and companies who are prepared to argue for the sort of government interventions that are needed to tackle sustainability challenges. As the reality of the business risks and opportunities associated with climate change have become clearer, so the business voices in support of action have become louder.

The Prince of Wales's CLG on Climate Change, for example, has made a number of very significant interventions in the UK and European Union (EU), offering strong support for climate policy measures (such as the 2008 EU climate and energy package) even when many other business associations were opposed to these measures.

Between 2007 and 2009, the CLG became one of the most influential business voices in the international climate debate, primarily through its series of hard-hitting, punchy statements in support of a strong, effective and equitable global climate deal. The most recent of these, the Copenhagen Communiqué, secured the support of more than 950 companies from more than 60 countries and was widely seen as the definitive progressive statement from the international business community ahead of the UN climate conference in Copenhagen (see CLG, 2009).

Just a decade ago, it would have seemed inconceivable to many that business groupings would be formed to advocate bold regulatory interventions from governments, in support of action on climate change. In a similar way, it might presently seem far-fetched to imagine that 950 companies would sign a statement calling for international regulatory action to protect biodiversity and ecosystem services. And yet, it is clear that there is a growing frustration at the failure of governments to keep up by introducing the regulatory frameworks required sustain biodiversity and ecosystem services. Earlier this year, for example, four retailers (Carrefour, Ikea, Kingfisher and Marks & Spencer) joined forces to call on the EU to ban the import of illegally sourced tropical timber. Their call recognized that labelling schemes such as the Forest Stewardship Council (FSC) can only go so far, and the business critical nature of their advocacy was clear.

At the launch of the new initiative, Mikael Ohlsson, CEO and president of the Ikea Group, said:

> *Wood is one of the most important raw materials for Ikea. It is an excellent choice from an environmental point of view, provided it comes from responsibly managed forests. We have worked for almost ten years to curb illegal logging, increasing the share of wood coming from respon-*

> *sibly managed forests. It is now important that decision makers take their responsibility, and act to introduce strong and efficient legislation.*

Sir Stuart Rose, chairman of Marks & Spencer, said:

> *Whilst we and our fellow members [of this initiative] are committed to responsible procurement, a lack of regulation means illegally harvested timber products can still enter the European market. Working together, [we] aim to send out a clear message that this is not acceptable.*

Conclusion

It would appear that 'biodiversity and ecosystem services' is on a similar issue trajectory to that travelled by climate change a decade before. As our scientific understanding of the functional role that species and ecosystems play in supporting humanity and the global economy increases, so the issue will rise up the political and corporate policy agendas.

What is different this time, however, is that so much of the analysis that is relevant to climate change is also relevant to that for biodiversity and ecosystem services. In his *Connecting the Dots: How Climate Change Transforms the Landscape of Risk and Opportunity*, for example, Burke (2009) describes the four 'pillars of prosperity' as climate, energy, food and water security. At least two of these can be addressed just as effectively through a 'biodiversity lens' as through a 'climate lens' but to address any of them properly, both approaches are needed.

And therein lies the real answer to the question 'is "biodiversity" the next "climate change" for business?', because many, including the author, would argue that we should perhaps stop seeing 'climate change' and 'biodiversity loss' as two separate problems but rather two related symptoms of the same underlying problem, namely the we are currently failing to manage the global economy in a way that respects environmental limits. Politicians will need a lot of support from business to tackle that one.

References

Bennett, C. (2009) 'What is a credible corporate response to climate change?', in F. Dodds, A. Higham and R. Sherman (eds) *Climate Change and Energy Insecurity – The Challenge for Peace, Security and Development*, Earthscan, London

Burke, T. (2009) *Connecting the Dots: How Climate Change Transforms the Landscape of Risk and Opportunity*, UN Global Compact, New York

CLG (Corporate Leaders Group) (2009) 'The Copenhagen Communiqué on Climate Change', www.copenhagencommunique.com

CPSL (Cambridge Programme for Sustainability Leadership) (2010) 'The Prince of Wales's Corporate Leaders Group on Climate Change (CLG)', www.cpsl.cam.ac.uk/leaders_group/clgcc.aspx

G8 (2010) *G8 Muskoka Declaration: Recovery and New Beginnings*, Muskoka, Canada, 25–26 June 2010, www.canadainternational.gc.ca/g8/index.aspx

Intergovernmental PBES (Platform on Biodiversity and Ecosystem Services) (2010) www.ipbes.net

IPCC (Intergovernmental Panel on Climate Change) (2007) *Climate Change 2007: The Physical Science Basis*, Contribution of Working Group 1 to the Fourth Assessment Report of the IPCC, Cambridge University Press, Cambridge

IUCN (International Union for Conservation of Nature) (2010) 'Red List overview', www.iucnredlist.org/about/red-list-overview

Jowit, J. (2010) 'Case for saving species more powerful than climate change', *The Guardian*, p1, available at www.guardian.co.uk/environment/2010/may/21/un-biodiversity-economic-report

Korten, D. (1996) *When Corporations Rule the World*, Berrett-Koehler, San Francisco

MEA (Millennium Ecosystem Assessment) (2005a) *Ecosystems and Human Well-being: Summary for Decision Makers*, World Resources Institute, Washington, DC

MEA (2005b) *Ecosystems and Human Well-being: Opportunities and Challenges for Business and Industry*, World Resources Institute, Washington, DC

Monbiot, G. (2000) *Captive State: The Corporate Takeover of Britain*, Macmillan, London

Rockström, J., Steffen, W., Noone, K., Persson, Å., Chapin, F. S. III, Lambin, E. F., Lenton, T. M., Scheffer, M., Folke, C., Schellnhuber, H. J., Nykvist, B., de Wit, C. A., Hughes, T., van der Leeuw, S., Rodhe, H., Sörlin, S., Snyder, P. K., Costanza, R., Svedin, U., Falkenmark, M., Karlberg, L., Corell, R. W., Fabry, V. J., Hansen, J., Walker, B., Liverman, D., Richardson, K., Crutzen, P. and Foley, J. A. (2009) 'A safe operating space for humanity', *Nature*, vol 461, no 7263, pp472–475

Stern, N. (2006) *The Stern Review on the Economics of Climate Change*, Cambridge University Press, Cambridge

TEEB (2010) *The Economics of Ecosystems and Biodiversity Report for Business – Executive Summary 2010*, Earthscan, London

UN (United Nations) (1992a) *United Nations Framework Convention on Climate Change*, available at http://unfccc.int/resource/docs

UN (1992b) *United Nations Convention on Biological Diversity*, available at www.cbd.int/convention/convention.shtml

21

A New Deal for Life

Jean Lemire

Nobody can claim to be above the fundamental laws of nature, since we are all products of nature. Life on the planet comes in many forms and the evolution of different living species seems to have favoured some species over others, whose role is often seen as secondary. Yet the more we study and understand the distinct role of each species in an ecosystem, the more we realize the vital importance of each of its elements. The smallest parts often support the largest and provide an essential function in the stability of different ecological systems which ensures the balance of life.

Humans sit at the top of the chain of life. Seemingly, the advancement of intelligence and awareness raised us to a prominent rank. But our position at the top of the pyramid of living does not necessarily exclude us from the ecosystem in which we belong, and does not justify a dominant claim to rise above the basic laws of nature. Humans also depend on the biodiversity of our ecosystem for survival, even if we do not always realize it.

Biodiversity is at the heart of ecosystems, and all life, ecological services and processes linking them, form a whole. Nature has been able to create, modify and maintain balance in these ecosystems that support life. It is from this vast pool of life that man finds its food, heals himself, builds his shelters and protects himself, and uses the abundant natural resources to create the energy he needs or the tools necessary for his development.

In our society we give a certain value to things. This is the very principle of our economic systems. Products or services are valued according to economic rules, and we convert the established value of a product or service by a monetary system that allows trade of these products of different value. This worldwide uniformity of barter has laid the groundwork for a new kind of global economy. However, in this complex business process, supply and demand often determines the value of products and services.

For example, without much regard for the processes of production, most countries provide a higher dollar value per litre of milk than a litre of gasoline, two universal products sold around the world. While milk requires minimal processing before being consumed, oil, on the other hand, is extracted from the bowels of the Earth in often difficult and costly conditions, transported over vast distances, refined and then sold on world markets before becoming available for local consumption. As the value of the products we consume is determined according to a market we created, it often falls on the architects of this economic system to establish an arbitrary value for these products or services. This simple rule inevitably leads to injustice, by simple discretionary choices that result in some form of economic aberration. Thus, the value given to the producer of milk, a cow, is disproportionately less than that given to the petroleum producer and their oil. But in our current and imperfect economic system, that overconsumption leads to skyrocketing demand for certain products; too often we underestimate the real value of ecosystem services that we take for granted.

For example, imagine a simple marsh that acts as a natural filter and supports a rich and diverse ecosystem. This small patch of wetlands and swamps has only a very small value in our market economies. It is often sold for a pittance, dried and converted into ordinary ground on which we build a new building, part of a housing complex in a new area annexed to a booming city. In this new area of housing, at great expense, we will build a wastewater processing plant by imitating artificially what nature was providing for free. Locally, nature has lost its precious natural functions. Most species living in marshes have been destroyed and the ecological services essential to this environment have been deliberately turned off.

This unique habitat, which offered a variety of plants, mature trees, frogs, insects, pollinators, birds, butterflies and small fish, required no maintenance as it was self-managed. Ecological services were numerous and all the surrounding ecosystems took advantage of the natural benefits of this small and uneventful swamp. Its transformation into a field without much ecological value has made it completely dependent on artificial new services created by human hands. Today, we invest to maintain its bright green grass; we use fertilizer and pesticides for its uniform green fleece on which someone will plant flowers and shrubs to imitate a certain nature. Paradoxically, in our view, the value of the 'new' field will, however, be much higher than that of the former marshes. The domestication of nature will bring more to its owner, according to an economic logic based on investment required for processing. Thus, by harnessing a parcel of wilderness and subdividing it into lots of living space, the new owner responds to a market demand we have created.

At first glance, there is nothing reprehensible about this simple transformation. We must respond to the growing demand for urban housing. After all, more than 50 per cent of the world's population now live in towns or cities and it is in developing countries that we are witnessing the fastest urban growth. But nature in the wild is losing more and more land and it is living species that pay the costs of this urban sprawl. The loss of natural habitats result in a decrease in biodiversity and now the bill for progress is accounted by important losses in plant and animal species that are constantly added to the International Union for Conservation of Nature (IUCN) Red List of Threatened Species.

The organized gathering of people under the umbrella of growing cities allows the establishment of services that make our lives easier, but this new phenomenon inexorably intensifies the increasingly important rupture with nature. In our well-organized and supplied cities, it no longer matters where our food is coming from. There is no need to hunt, fish or grow crops, the supermarket has everything one could ask for and more. It is also not necessary to walk to a source of drinking water and transport this vital commodity to our shelter, the water now comes from the water system and we just turn a valve and drain the blue gold. Such appearance of abundance no longer allows the identification of the source or its fluctuations, which formerly limited our natural ways of consumption.

The search for fuel is also a thing of the past. The fire for cooking comes from the stove and fuel for our cars is accessible anywhere, anytime. Our waste is loaded into trucks and takes the road to unknown and out-of-sight destinations. The sedentary lifestyle has changed the lives of our children and the computer screen has replaced group activities. In our urban islands, for some children, the mouse more appropriately refers to the computer instead of the small rodent and it might be time to teach some people that 'fish and chips' is not a fish species from northern Europe!

Nevertheless, we must admit we like this new world of facilities. In our modern cities where everything is easily available, it is no longer necessary to provide the appropriate effort to obtain the fruits of our labour. However, although one can not be against progress and modernity, we must revise our models to find some form of harmony and respect for nature. Our current social system created a deficit far too important to continue without major changes. Already, each year, our consumption of resources in the world surpasses what the Earth happens to produce. This lack of resources is likely to increase if nothing is done, especially with high population growth rates. We must rethink the way we consume and especially our production processes. The necessary transitional phase to sustainable development policies is our best investment for the future of humanity.

The examples are numerous and show how we have distanced ourselves from nature in its wild state. One solution to counter this imbalance lies in education. We must redefine our relationship with nature. In our scheme of values, the reconstruction of an environment with a fake 'natural' look too often prevails over wild nature. On this point, we have it all wrong and we must review the actual valuation of nature in its raw state.

What is the real value of bees that are responsible for pollination of flowers and crops in the current economic system? And that of coral reefs and mangrove forests that protect our coasts? Who would give a monetary value to bacteria and other microorganisms that play a vital role in our lives? Based on this principle, by which domesticated nature is often attributed a greater value than nature in the wild, we influence directly our measures of appreciation – and therefore our economic relation – with different forms of life on the planet. In our analysis of profitability, environmental services and so-called secondary species are deliberately ignored in the economic balance, which often leads us to make irrational decisions. In ignorance – or worse, in indifference – we encourage a failing system and pursue an organized and planned destruction of nature, without realizing the immense damage done to ecosystems that form the best insurance policy for our own survival.

More than ever before, we are destroying life. Today, living species are disappearing at a rate never equalled in the history of the Earth. The coexistence between humans and other life forms that inhabit this planet is deeply imbalanced and the law of the strongest prevails, forever sacrificing species that are now out of the living picture of this planet. We must, without hesitation, quickly review our practices to allow other forms of life to flourish. If modern humanity is to keep going at the frantic pace of its development, it will have to take into account the vital concept of sustainability if it wants to successfully face its pressing and inevitable population growth. Nature can not withstand a population of 9 billion people by 2050 if nothing is done to rethink our sharing of natural resources.

In our search for solutions, we must understand that the overall challenge goes far beyond local concerns. The decline in bee populations, the disappearance of amphibians or the alarming planned deforestation will have important consequences on our lives, no matter where we live. The imbalance of major ecosystems may lead to costs that are difficult to assess in our global economy as a whole. Insects are responsible for pollination of our crops, and frogs, besides controlling invasions of insects carrying diseases, contribute to the ever-growing evolution of medicine and pharmacology. The removal of ancestral forests, to the detriment of monocultures, deprive us of essential ecological services. Vast forests are the lungs of the planet and they participate in the wider global climate system by absorbing large amounts of carbon dioxide. In the midst of a global climate crisis, we must remember their value and recognize the beneficial and essential contributions of these large forests at the global scale. We will have to respond quickly and effectively to limit these violations to the code of life if we want to preserve the overall balance of biodiversity. This fundamental rethinking of the pact between mankind and other living species has become mandatory to ensure the survival of different life forms on this planet, including ours.

So, what should we do to participate in this wind of change to recover from a situation that is doomed for failure? Of course, we can put the blame on the wealthy, protest against globalization, decry consumerism that depletes natural resources of the planet for the benefit of a minority of people and make a scene about the big profits of big companies, banks and other managers of the powers that control and set the value of products and services. But it will take more than opposition, speeches and demonstrations to turn a global situation around. Despite the undeniable weight of the arguments against consumption society and the undeniable responsibility of its main economic actors, it would be futile and unrealistic to imagine a radical change in the rules of our system without the active participation of those who control this system. At the same time, it would be unrealistic not to include those who feed this system – us, the consumers. In front of this great challenge we all face, the exclusion of certain actors of our society is certainly not an option.

Citizens, governments, large corporations, people of all countries and all backgrounds, rich and poor, we must all contribute, in our way, to the implementation of a new pact between humans and nature. Although it might be utopian to fight or completely abandon consumption, we have a duty to review our ways of life, because, whatever people may say, the problem is not simply economic. It is also and above all, philosophical. The revolution must go through the attribution of real and tangible services provided by nature, its species and a greater respect for all life forms involved in the equilibrium of ecosystems.

Nobody wants the disappearance of a species. Nobody feels threatened by a bird, a fish or a simple plant. But consciously, we now accept the final demise of many species that are part of our environment, as if we were above the laws of nature. As if the ecosystems that support all life forms that took millennia to create and to achieve balance no longer needed all their components in order to exist. As if a single species could now control life.

More than ever, we see ourselves outside nature and thus we believe we control everything. However, we are like all other creatures, we come from nature and we are a part of it. Despite the appearance of our civilization, we are no more separated from nature than animals, plants, bacteria and other forms of life in our ecosystems. We must remember the concept of interdependence of all life forms, including ours.

Because we developed intelligence and consciousness – that distinguish us from many forms of life – we apparently raised ourselves above the laws of nature. Paradoxically, this is this same consciousness that leads us to judge our actions. However, to reduce the moral weight of our own judgement, we have developed a range of socially acceptable behaviours that helps us keep a clear conscience. It is the very characteristic of intelligence: the knowledge of how to play with situations and their analyses to create an accommodating and comfortable argument to justify the means of its ambitions. This is indeed the normal logic of a majority of politicians. Policies stem from a stream of short-term thinking, in favour of current public opinion, for the praise and the favour of an electorate that, in a democratic system, has the duty to assess and judge those in power.

But how can we expect a long-term vision from our leaders if we indulge ourselves in our minimum clear conscience? Small actions, while important, are only the prelude to the real desired change. But they must evolve, like life on this planet, to achieve the desired balance. We are children of change. We must continue to transform our vision to succeed in stretching our vision towards the future, to ensure a better balance in our lives, to rediscover our relationship with nature and invest in a future full of promise and respect for life in all its forms.

To counter the ignorance and indifference, we must raise awareness, educate, demonstrate and explain about fragility. The universal accessibility of social media has created a new world without borders in which information flows easily between people and from one country to another. Thus, we can rejoice in the fact that children are now experiencing the problems of deforestation in the Amazon, but at the same time, they are unaware of their wooded district. Curiously, the rupture with nature has provoked a real passion for nature, but theory currently prevails over practice. We claim the preservation of forests, the protection of water streams and the reduction of emission of greenhouse gases (GHGs), giving the excuse that we recycle and turn the tap off while brushing our teeth. We must do more. Now that we know the major issues for the future, we must acknowledge, humbly, that we are a major component in the great chain of life, where everything is connected.

Indigenous peoples from all regions of the world speak, each in their own way, about this notion of interdependence with nature. They understand that we are only a fraction of a whole, an element of a community of lives, or rather a spark of life among the multitude of organisms like ourselves. Therefore, indigenous peoples attach much importance to nature. They understand that the threats to nature are also dangers for them, and for all elements of nature.

Biodiversity and its decline should ironically affect and lay the paths for the future, reconciling ecological and economic thinking. Maybe we had to push ourselves to the brink of the precipice to see the full extent of our carelessness? Despite an ever-increasing rupture between man and nature, hope may be revived in the simple recognition of beauty. It is our duty, during this period of change and transition of societal values, to put the simple beauty of nature at the forefront and the preciousness of life and its fragility. Any reasonable person can not wish for the vanishing of life. We must communicate this beauty, show it and continue to spread the alarm expressed by threats to plants, fishes or birds. Often, the melodious chirping of the bird happens to touch souls more than the green protesting sermons, and can then initiate reflection. Nature, in all its beauty and its fragility, will reconcile the minds of ecologists and economists. In the near future, it is hoped that disrespect for life will not be acceptable, because we do not want the new values of our societies to be inspired by our past mistakes.

The evolution of the 'green movement' and appropriation of some form of social ecology are probably the sources of the first stuttering of a new society. Governance will have no choice but to build on these new values, otherwise it will not survive. We must recognize the phenomenal results of the environmental movement in recent decades. Methods can be discussed, sometimes radical or controversial, to awaken the conscience of people. Despite the blunders and mistakes along the way, history will remember the victory of current ecological thinking and its influence on power as a turning point in the evolution of our societies. The ecological thought, if it manages to take root deeply in the heart of the masses, will breathe this essential and most desired wind of change.

In this transition period, we should not believe that the progressive installation of new values in our societies will be 'the solution' to the problems of the planet, but it will slow the process of current erosion and will give time for a permanent, healthy change of the mentality of our society to take root. Our youth, who will be in power tomorrow, will learn from the past and assume its responsibilities for the future of life. Hopefully, this generation will introduce the basic principles of sustainable development into its decisions, and hopefully this new attitude will change forever the rules of governance on this planet by assigning to ecological services the real value they deserve. This new aspiration, led by the desire of younger generations, remains the greatest source of hope for tomorrow.

While we have not bequeathed a healthy planet to future generations, we may have managed to instil in the young leaders of tomorrow the core values of respect for a better coexistence between humans and other life forms that share this unique planet. The hope is real and belongs to those who support and feed the actions that lead to real change. If the trend continues and this common desire to protect life definitively settles, our children will be able to enjoy the fruits of our scarce seeds: the seeds of consciousness deeply buried in the bowels of a new Earth, sown by the hands of man, and blown by a wind of change and hope for future generations.

Index